BIOSYSTEMATIC MONOGRAPH
OF THE
GENUS *CUCUMIS*
(CUCURBITACEAE)

Botanical Identification
of
Cucumbers and Melons

BIOSYSTEMATIC MONOGRAPH

OF THE

GENUS *CUCUMIS*

(CUCURBITACEAE)

Botanical Identification

of

Cucumbers and Melons

Joseph H. Kirkbride, Jr.

U.S. Department of Agriculture
Agricultural Research Service
Systematic Botany and Mycology Laboratory
Beltsville, Maryland 20705-2350
U.S.A.

Parkway Publishers
Boone, North Carolina
1993

Parkway Publishers
Box 3678
Boone, North Carolina 28607 USA

Library of Congress Cataloging-in-Publication Data

Kirkbride, Joseph H., 1943-
    Biosystematic monograph of the genus Cucumis (Cucurbitaceae) :
botanical identification of cucumbers and melons / Joseph H.
Kirkbride, Jr.
        p.   cm.
    Includes bibliographical references (p.   ) and indexes.
    ISBN 0-9635752-0-1
    1. Cucumis--Classification. 2. Cucumis--Identification.
I. Title.
QK495.C96K57 1993
583'.46--dc20   92-82126

Purchase of this publication licenses use of the program INTKEY on only one microcomputer at any one time, and only with the data in this package. Later versions of the program may also be used with these data (but there is no guarantee that later versions will be compatible with the data). Requests for program support must be directed to the author of this monograph, not the authors of the program. Commonwealth Scientific and Industrial Research Organisation (CSIRO), Australia, or its employees, Parkway Publishers, or the author shall not be liable for any loss, damage, or injury (including without limitation any loss of profit, indirect, consequential, or incidental loss, damage, or injury) arising from use of the program. The files DELTA.USE and DELTA.REG contain information on obtaining program upgrades from the Internet or from the program authors, and on registering INTKEY and related programs for use with other data sets.

CONTENTS

TABLES AND ILLUSTRATIONS

PREFACE

The amount of information concerning the biological world is immense and growing everyday. The key to accessing that information is the scientific names of organisms. Systematics has a central role in data organization. Its adherents synthesize all available data for organisms, look for new data, and then develop the best possible classifications supplying the scientific names that allow others to accurately access all information for organisms. The objectives of this book are to facilitate the identification of cucumbers, melons, and their wild relatives and, following identification, to present the correct scientific names. Just for commonly cultivated melons, *Cucumis melo* Linnaeus, there are 529 scientific names, only three of which should be used.

The world's population is sustained at its current levels by modern agriculture. The basis of that agriculture is highly productive cultivars that have been tremendously improved through application of modern breeding and improvement technology utilizing germplasm available from the worldwide network of germplasm systems. Accurate and productive access to germplasm is through the scientific names of organisms. This book will insure the correct application of scientific names for melons and cucumbers, and thereby contribute to use of their germplasm in breeding and improvement programs.

The information in this book is being presented in two formats, the traditional printed page and microcomputer diskette. The printed portion presents data in the customary fixed format. The microcomputer diskette furnishes data in an electronic format accessible via MS-DOS computers. Using appropriate software, this data can be accessed in a myriad of ways allowing users to tailor the data to their needs. This book heralds the opening of an electronic age for systematic data.

I am grateful to Charles Jeffrey, Royal Botanic Gardens, Kew, for his generous help, cooperation, and insights into the systematics of the Cucurbitaceae. Dr. Mats Thulin allowed access to his manuscripts on *Cucumis* for the *Flora of Somalia* and provided other data on *Cucumis* in Somalia.

The following institutions loaned specimens for this study (acronyms according to Holmgren et al., 1990): BH, BLAT, BM, BOL, BR, C, CAL, CANB, CM, E, GAT, HUJ, K, KUN, L, LE, LISC, LISU, LIV, LMU, LY, M, MEL, MO, MPU, NSW, PE, PRE, S, SRGH, TI, UPS, W, and the Vegetable Research Institute, Beijing. The study could not have been done without their support and patience. In addition visits were made to the Royal Botanic Gardens, Kew, the Natural History Museum (formerly British Museum [Natural History]), London, the Linnean Society of London, London, the Muséum National d'Histoire Naturelle, Paris, and the New York Botanical Garden, Bronx, during which their historic collections, libraries, and archives were studied. The Institute for Horticultural Plant Breeding, Wageningen, was also visited, and its staff was an important source of information on the biology and cultivation of *Cucumis*. The collections and resources of the National Agricultural Library, Beltsville, the U.S. National Arboretum, Washington, D.C., and the Smithsonian Institution, Washington, D.C., were regularly consulted.

I thank Dr. Charles E. Jarvis, Linnean Plant Name Typification Project, Natural History Museum (formerly British Museum [Natural History]), London, for his essential advice and assistance with the typification of Linnean epithets. The Botanical Research Institute, Pretoria, supplied a printout of *Cucumis* data in PRECIS, facilitating the interpretation of collection data from southern Africa. *Cucumis* germplasm supplied by the U.S. Department of Agriculture, Agricultural Research Service, National Plant Germplasm System, and the Institute for Horticultural Plant Breeding, Wageningen, was grown out and supplied an organismal glimpse of *Cucumis*. The staff of the Institut de Botanique, Montpellier, sent copies of the *Index seminum, horti regii botanici Monspeliensis* published by Alire Delile (Raffeneau) in 1843 and 1847 and supplied information on all issues from 1843 to 1859.

Carole A. Ritchie, Laurel, Maryland, and Judith R. Purman, Owings Mill, Maryland, with patience and skill transferred specimen data into a microcomputer database. Dr. John H. Wiersema, U.S. Department of Agriculture, Agricultural Research Service, Systematic Botany and Mycology Laboratory, was an unfailing source of nomenclatural information and advice. Karen Parker, Bowie, Maryland, prepared the illustrations that grace and clarify this study.

I thank Drs. Rupert C. Barneby, the New York Botanical Garden, Bronx, Paul A. Fryxell, USDA, ARS, Agronomy Field Laboratory, College Station, Texas, Charles Jeffrey, Royal Botanic Gardens, Kew, M. Thulin, Botanical Museum, Uppsala University, Uppsala, Richard W. Robinson, New York Agricultural Experiment Station, Cornell University, Geneva, New York, and John J. Wurdack, Smithsonian Institution, Washington, D.C., for their constructive reviews of the manuscript. Dr. Barneby also supplied illuminating information on botanical Latin.

Joseph H. Kirkbride, Jr.
Beltsville, Maryland

INTRODUCTION

The genus *Cucumis* Linnaeus (Cucurbitaceae) includes two major commercial vegetable crops, cucumbers and melons, and two minor ones, the West Indian Gherkin and the Kiwano. The former are also common elements in home gardens (Neier and Marr, 1988). Both the cucumber and the melon were known from the earliest times (Pangalo, 1929d; Manniche, 1989). Each has distinct names in ancient Egyptian and Coptic, and both appeared in Theban tomb paintings. They were consumed as part of the Egyptian diet, and were utilized as medicinals. The Greeks and Romans also cultivated them (Pangalo, 1929d; Rackham, 1959; Jones, 1951; Manniche, 1989), but the Romans did not have distinct names for them. They referred to melons as 'ripe cucumbers' (Manniche, 1989) sometimes making it difficult to know to which they were referring (Rackham, 1959; Jones, 1951). During the era of the great European herbals in the 16th and 17th centuries (Pangalo, 1929d; Davis and Heywood, 1973; Woodward, 1964), both were presented as important vegetables and medicinals.

Today, their importance as vegetables and fruits continues both commercially (Anonymous, 1990) and domestically (Neier and Marr, 1988). *Cucumis* has long been known to be susceptible to a number of devastating fungal, bacterial, viral, and insect diseases (Whitaker and Davis, 1962). These diseases reduce both the productivity and quality of *Cucumis* crops. Intensive investigations are underway to discover resistance to many of these diseases (Thomas, 1989). The basis of these studies is the world's stock of *Cucumis* germplasm. Of 76 *Cucumis* germplasm samples examined in 1976, 25 or 33% were incorrectly identified (Leeuwen and Nijs, 1980), and others were distributed under archaic names.

Each monographic study of *Cucumis*, including this one, has been induced by significant increases in information about the genus and desires to order that information into patterns more reflective of those found in nature. In the nineteenth century, for the first systematic monograph of *Cucumis* (Naudin, 1859) there were at least 150 extant herbarium collections of the genus most of which were available, and for the second systematic monograph of *Cucumis* (Cogniaux, 1881) there were approximately 350 collections available. In the twentieth century, for the third systematic monograph of *Cucumis* (Cogniaux and Harms, 1924) there were more than a thousand collections available, and for this biosystematic monograph of the genus 3,881 collections were studied in depth. Thus, there has been a 25-fold increase in available data from herbarium specimens between the first monograph and this biosystematic study, an 11-fold increase between the second monograph and this study, and a 3.8-fold increase between the third monograph and this study.

In the third systematic monograph of *Cucumis* (Cogniaux and Harms, 1924), 38 species of *Cucumis* were recognized. The explosion of *Cucumis* information, especially biosystematic data, necessitated a re-evalutaion of those species. In the present study, 16 of those species are considered to be synonymous with pre-existing species, one is accepted as a species of *Cucumella* Chiovenda, and 4 are indicated to be synonymous with *Oreosyce africana* Hooker f. In addition, 15 more species, not then known, are accepted in the present study.

The objectives of this study are to supply the necessary information for the correct botanical identification of plants of the genus *Cucumis* and for the evaluation of their botanical names. To achieve these ends the following are presented: the taxonomic history, morphology,

1

and biosystematic data of *Cucumis* are reviewed with an emphasis on those data relevant to its systematics. *Cucumis* itself and all infrageneric taxa are fully described with keys for the identification of all infrageneric taxa. Names whose placement is uncertain and names excluded from *Cucumis* are listed. The synonyms of *C. melo* Linnaeus and *C. sativus* Linnaeus are given in appendixes. The characters used for the collection and storage of morphological, cytological, and distributional data are also listed in an appendix to elucidate the data formats utilized in the preparation of the identification aids. An alphabetical index of the exsiccatae studied in-depth is included to facilitate the identification of duplicate specimens. Accompanying this publication is a 5¼ inch DOS diskette with the morphological-cytological-distributional and detailed collection databases. Using these databases, the data can be manipulated and customized to fit specific needs. This study is one step in the long process leading to total understanding of the genus *Cucumis*, and should stimulate further studies clarifying the status of its little known species.

TAXONOMIC HISTORY

Linnaeus presented the genus *Cucumis* in *Species plantarum* (1753) and in *Genera plantarum*, ed. 5 (1754). He accepted seven species in the genus, all of which were cultivated or economically useful; three are maintained in this study, one is synonymized, and three have been transferred to other genera in which they are accepted. In subsequent years, Linnaeus named two additional species, one of which is accepted and the other reduced to synonymy. Miller (1754) created the genus *Melo*, but he neither described any species in it nor transferred any species to it. In the remainder of the 18th century, numerous new names were published for the utilized species, but almost all of them have been relegated to synonymy.

In the 19th century, there were two major trends in the taxonomy of *Cucumis*, one related to the era of overseas exploration and expansion by Europeans and the other to the rapid advancement of agriculture and crop improvement in Europe. The former is epitomized by the monographic studies of *Cucumis* and the family Cucurbitaceae by Naudin (1859) and Cogniaux (1881). Naudin monographed the genus *Cucumis*, and accepted 13 species arranged in two informal groups, one with spiny, muricate, or at least tuberculate fruits, the other with unarmed, pubescent, or glabrous fruits. These groups were described as natural, but subject to modification as further *Cucumis* species became known. Naudin's work was based almost exclusively on living plants grown in the gardens of the Muséum d'Histoire Naturelle, Paris. He grew all available species over various years, prepared numerous herbarium specimens, and observed the variation among individuals, between generations, and from year to year. Extensive notes and watercolor drawings were made to preserve data for ephemeral characters, such as fruit color, not observable on herbarium specimens, and these materials were deposited in the herbarium and library of the Muséum d'Histoire Naturelle. His species and their delimitations were biologically oriented and very close to those that would be obtained through application of the biological species concept (Davis and Heywood, 1973; Stace, 1980). Cultivated types were placed together with their wild relatives in single species, and cultivar variation was handled by the recognition of infraspecific taxa. He was the first to document hybridization among some *Cucumis* species (Naudin, 1862, 1866).

Cogniaux was the taxonomic specialist on the family Cucurbitaceae in the closing decades of the 19th century. He prepared a world monograph of the entire family for Alphonse de Candolle's *Monographiae phanerogamarum* (Cogniaux, 1881), and so monographed *Cucumis*. He accepted 26 species in the genus arranged in 10 informal species groups identified by the species numbers of the included taxa:

3

I. Fruit smooth, glabrous, or pubescent.
    A. Ovary densely pubescent or villous.   Species Group 1
    B. Ovary hispid or aculeate.   Species Group 2
II. Fruit echinate, spiny, or tuberculate.
    A. Stems procumbent, with tendrils.
        1. Perennial herbs.   Species Group 3
        2. Annual herbs.
            a. Stems with slender, fragile vestiture of hispid, pilose, or long setose
                pubescence.   Species Group 4
            b. Stems scabrous, shortly white-pilose, and thick-aculeate. Species Group 5
    B. Stems erect, without tendrils   Species Group 6

These groups and their descriptions correspond exactly to the first couplets and the species arrangement in his key. His treatment of *Cucumis* is very different from Naudin's. He had a great many more herbarium specimens available, and he worked principally from them. Consequently, his species delimitations are narrower than Naudin's and less biological in their nature.

The English colonial floras (Harvey and Sonder, 1862; Hooker, 1871; Clarke, 1879; Hiern, 1898) also contained significant contributions to the genus *Cucumis* in the last half of the 19th century. They documented the rich collections arriving from Africa and India, and clarified the distributions of many wild species of *Cucumis*, especially in Africa.

The other 19th century trend in the taxonomy of *Cucumis* is best characterized by the work of Alefeld (1866) and Harz (1885) in Germany. They focused upon the fruits of *C. melo* and *C. sativus* and established infraspecific botanical epithets for each of the agricultural cultivars known to them. They grouped together wild relatives and cultivar variation in single species as Naudin and Cogniaux had, but they attempted formally to recognize cultivar variation through numerous infraspecific epithets of the same rank.

At the beginning of the 20th century, Harms and Cogniaux (1916, 1924) began a world monograph of the Cucurbitaceae in Engler's *Das Pflanzenreich* based on Cogniaux's previous treatment (Cogniaux, 1881), but they published only the first three tribes. *Cucumis* appeared in the second installment (Cogniaux and Harms, 1924) and included 38 species. Many of the descriptions are almost identical to those published by Cogniaux in 1881, and the key is also very similar. They used exactly the same beginning couplets in their key and ordered the species in the same way as Cogniaux had in 1881, with the intercalation of species discovered after 1881. They did not, however, refer to species groups in any way. They followed Cogniaux's previous work very closely, the main differences being the addition of taxa published since 1881 and citation of herbarium collections made since then.

The botanical naming of cultivars with emphasis on fruit types thereafter shifted from Germany to the USSR. A rarely cited, extended series of mainly Russian publications (Pangalo, 1925, 1928a, 1928b, 1929a, 1929b, 1929c, 1929d, 1929e, 1933a, 1933b, 1934, 1937, 1950; Gabaev, 1929, 1932, 1933a, 1933b; Kozuchov, 1930; Grebenscikov, 1953; Filov, 1960) has resulted from this work. At first, the Russian workers established their new infraspecific taxa in the Linnean hierarchy of taxonomic ranks. In later publications, they expressed dissatisfaction with the Linnean hierarchy as a framework for presentation of the variation found within cultivated members of *Cucumis*, and they began to use an assortment of ranks defined for the treatment of cultivated plants (Jirásek, 1961). Grebenscikov (1953) and Filov (1960) attempted to present complete systems of classification using these ranks for the cucumber and melon, respectively. Pangalo (1950) resurrected the genus *Melo* and moved the members of *Cucumis melo* into it. Some of the principal cultivar groups were treated as species, and an extensive infraspecific hierarchy was created for Cultivars. Four sections and two series in the genus were also created. They all fall into synonymy under *Cucumis* ser. *Melo*. No other author has adopted this scheme for the melons.

Since Cogniaux and Harms (1924), a large number of floristic treatments have appeared for *Cucumis* (Davy, 1926; Hassib, 1938; Chakravarty, 1946, 1959, 1961, 1966, 1968; Berhaut, 1954, 1967, 1975; Fernandes and Fernandes, 1962, 1970; Meeuse, 1962; Keraudren, 1967, 1968; Jeffrey, 1967; Roessler, 1968; Kerauden-Aymonin, 1975; Fernandes, 1976; Andersen, 1977; Feinbrun-Dothan, 1978; Halliday and Jeffrey, 1978; Chakravarty and Jeffrey, 1980; Telford, 1982; Reekmans, 1983; Nazimuddin and Naqvi, 1984). Their focus was on wild species with superficial treatments of the cultivated ones, and none of them referred to the infrageneric classification. Those of Meeuse (1962) and Jeffrey (1967) covered the largest number of taxa. The species treatments and delimitations varied from work to work and are difficult to compare.

The circumscription of *Cucumis* has been stable since Linnaeus (1753) except for the unsuccessful attempts by Miller (1754) and Pangalo (1950) to separate the melons as a distinct genus, *Melo*. Chiovenda (1929) described three specimens from Somalia as the new genus *Cucumella*. Jeffrey (1962) took up the genus and expanded it by transferring five epithets into it, one from *Cucumis*, *C. asper* Cogniaux, one from *Hymenosicyos* Chiovenda, two from *Kedrostis* Medikus, and one from *Oreosyce* Hooker filius. The principal difference between *Cucumis* and *Cucumella* is that their anther thecae are either sigmoid or straight, respectively. Following Jeffrey, authors of the Cucurbitaceae in African floras (Keraudren, 1966, 1967, 1968; Jeffrey, 1967, 1975; Roessler, 1968; Fernandes and Fernandes, 1970; Kerauden-Aymonin, 1975) have accepted *Cucumella*. Throughout Jeffrey's (1962, 1980a, 1990) studies of the subfamilial classification of the Cucurbitaceae, he has maintained *Cucumella* as an independent genus closely related to *Cucumis*. Recently Jeffrey (pers. comm.) and Thulin (pers. comm.) have expressed opinions that *Cucumella* may be paraphyletic, and that it may be best to include its present species in *Cucumis*. I have chosen to use the widely accepted circumscription of *Cucumis*, excluding *Cucumella*, pending further evidence clarifying the circumscription and status of *Cucumella*.

In this monograph, I have separated the nomenclature of wild organisms from that of cultivated organisms, as has been done in the current international codes of nomenclature for plants (Greuter et al., 1988; Brickell et al., 1980). The creation of botanical taxa for each of the myriad variations found in cultivated plants masks the genetic unity of species by emphasizing their minor differences, many of which are ephemeral under natural conditions, over their similarities. Those traits desired by man in his cultivated plants, when introduced into botanical taxonomy, produce a haze of confusion around taxa and lead to a false sense of genetic distinctness (Doebley and Iltis, 1980; Iltis and Doebley, 1980). However, those same traits are key to understanding the variation within cultivated plants and framing their systems of cultivated nomenclature. In *Cucumis*, fruit characters are critical to understanding the delimitation of most specific taxa, including the cultivated ones, but I have deliberately not used those characters within *C. melo* and *C. sativus* when defining infraspecific taxa.

MORPHOLOGY

The morphology of *Cucumis* is remarkably stable for some characteristics of particular organs, but for other characteristics of the same organs the morphology can be highly variable. Some organs have more stable characteristics than others, and characteristics that are stable in some taxa may be highly variable in others. The use of characters must always be considered in the context of the organ and the taxon. The character traits desired by man in the cultivated forms of *Cucumis* have been so manipulated and maintained against natural selection that the range of variation in those traits far surpasses that found in the remainder of the genus. So for many characters, there is a normal range of variation as commonly encountered under natural conditions and the extreme limits of variation, very rarely found under natural conditions, that occur in cultivated organisms. Comparison of specimens of the same taxon collected under natural conditions with cultivated plants or with specimens prepared from them sometimes reveals character differences complicating the identification of cultivated *Cucumis*. Manipulation of environmental conditions of cultivation has not been able to induce some of the same characteristics found in organisms under natural conditions (Leeuwen and Nijs, 1982). Consequently the botanical identification of cultivated *Cucumis* specimens can be very difficult.

The numbers enclosed by parentheses in the following discussions refer to the characters in Appendix 3, DELTA (DEescription Language for Taxonomy) Character List for the Species of *Cucumis*.

Habit

Collection labels very seldom describe the habit (1); frequently it must be inferred from the characteristics of the specimens. Most species produce soft, herbaceous stems and branches. Taxa with soft stems and not described by their collectors as climbing were characterized as herbs, and those with woody stems as subshrubs. The type of root system, whether woody or not (3), cannot be determined for the majority of specimens since they lack roots. This has been used to separate *C. trigonus*, with a woody taproot, from *C. melo* without one. Further studies are needed to determine whether the development of a woody taproot is facultative in some *Cucumis* species.

The surface ornamentation of stems is variable. Most species have pubescence on their stems (14), and two, *C. aculeatus* and *C. ficifolius*, have aculei (13). Jackson's (1928) definition for aculeus, "a sharp epidermal emergence, a prickle," which agrees with that of Payne (1978), has been accepted here. Aculei are easily separated from hairs by their larger size and whitish or cream color. Their presence or absence is a stable character, and usually they are found on more than one organ. Consequently, it is an important character for the identification of many species, and so has been considered separately from pubescence to facilitate key generation (13, 21, 36, 56, 135, 145, 201). *Cucumis sacleuxii* has a distinct hair type on its stems and other organs that is referred to as 'breakaway' hairs (15, 24, 39, 128, 193). They are the *non-glandular, multicellular, conical, simple* type of Inamdar and Gangadhara (1975) and Inamdar et al. (1990). They consist of a multicellular foot and a uniseriate apex with the cell walls of the foot much thinner than those of the apex. The cells of the foot are easily ruptured, separating the

7

apex from the plant. Possibly this is a defense mechanism against predation. Internode length is one of the characters affected by cultivation (18). On some cultivated specimens, it reaches exaggerated sizes, which are always reported in parentheses as extreme ranges.

Even though *Cucumis* species have tendrils (53-58), they seldom, apparently, support themselves by climbing or attaching to something erect. Their stems and branches lie across the surface of the ground without rooting. Each node has a single leaf and tendril, except *C. humifructus* which has a fascicle of 5-8 tendrils and *C. rigidus* which lacks them.

Subterranean tubers have been described on the roots of a single species, *C. kalahariensis* (6-10). According to its collectors, the tubers are edible and are eaten by native peoples of the Kalahari desert of southern Africa.

Leaves

Leaves are the most difficult organs to characterize because they are highly variable, especially in size, and are affected by cultivation. I have seen leaves of *C. sativus* in greenhouses in the Netherlands that were more than 50 cm wide. The petiole, in general, is not very useful in delimiting taxa except for its hair type (22). All species, except three, have a single, uniform pubescence type on their petioles. *Cucumis sagittatus* and *C. thulinianus* have two pubescence types uniformly mixed over the entire petiole, and *C. myriocarpus* has three types separated into distinct zones on the petiole, retrorse-strigose at base, hirsute in the middle, and antrorse-strigose at apex.

Leaf blade lobation is variable (25). The degree of lobation is indicated by the terms used. Trilobate, quinquelobate, and sexlobate usually have the sinuses reaching from the apex less than half way to the base of the blade, and 3-palmately or 5-palmately lobed indicates that the sinuses generally are recessed halfway or more from the apex to the base. Blades of most taxa tend to be either less lobed or more lobed, but this is difficult to quantify. The leaf blades of some taxa are pinnatifid, i.e. with short lateral secondary lobes (42). Generally the number and position of the lateral lobes are highly variable. The subgen. *Cucumis* has notably distinct leaves, but their peculiarity is difficult to describe. They are very shallowly lobate, and the lobes are generally triangular.

Tendrils

The tendrils are in general uniform (53-58). They vary greatly in length (56) depending upon their age and whether they have twined about something. *Cucumis insignis* was described from a single specimen the tendrils of which are either simple, as in all other *Cucumis* species, or they are bifid (55). Further data are needed to fully understand this phenomenon and evaluate its taxonomic usefulness.

Inflorescences and Flowers

Most species are monoecious. *Cucumis humifructus* has only androgynous inflorescences, i.e., inflorescences with both female and male flowers and the female flower below the male ones, and *C. metuliferus* has mainly unisexual inflorescences and a few gynecandrous ones, i.e., inflorescences with both female and male flowers and the female flower above the male ones (59). The significance of these inflorescence types is difficult to interpret without more data. Cultivated forms of *C. sativus* have many types of sexual expression: Monoecious, gynoecious, andromonoecious, androdiecious, hermaphroditic, and others (J.E. Staub, pers. comm.). It is difficult to identify these in dried specimens.

Male Inflorescences and Flowers

Male inflorescences are often multiflowered (63) and are rarely branched (62). When the inflorescences are branched, the male flowers are always pedicellate. The pedicel of the flower is subtended by a small bract (70-75) only in *C. heptadactylus*. Most male inflorescences are reduced to a single flower supported by a stalk. Because branched inflorescences have pedicellate flowers and the only bracteolate species has the bracts at the base of the pedicel, the stalk of 1-flowered inflorescences is here considered to be a pedicel, not a peduncle.

Male flowers are very uniform throughout the genus. There is variation in size among some species, and the cultivated forms exhibit the largest flowers. The stamens are typically cucurbitaceous, two with bithecate anthers and one with a unithecate anther. The anthers are shortly pubescent on the edges, and the hairs of one anther frequently intertwine with those of the next, so that they resist separation and form a tube. Perhaps this promotes the transfer of pollen on to pollinators. The thecae themselves are thrice folded into a sigmoid shape which distinguishes *Cucumis* from *Cucumella*, the most closed related genus (Jeffrey, 1976a, 1980a, 1990). At the base of the hypanthium there is a cylindrical disc, rarely reduced to three papilliform structures (110-112). There is no other apparent structure in the position occupied by the gynoecium in the female flower.

Female Inflorescences and Flowers

Female inflorescences are almost all one-flowered, but a few consist of a fascicle of flowers (113). Uniflowered inflorescences are interpreted like male ones as sessile with a pedicellate flower.

Again the general pattern of the female flowers is very uniform throughout the genus, and size varies as in the male, with the largest ones in the cultivated forms. The shape of the pedicel, linear or dilated upward from base to apex, is a very useful character (120). Sometimes it is difficult to determine before fruit development, but the fruit pedicel always has the same shape as that of the female flower. The hypanthium, consisting of fused sepals, petals, and stamens, is hour-glass-shaped with the lower portion, including the constriction, adnate to the ovary. The surface ornamentation of this area generally is the same as that found on the fruit (125, 126). The most striking difference between the two is found in *C. sativus* and *C. rostratus*, in which the lower portion of the hypanthium is aculeate but the fruit is unarmed. Apparently the aculei are overwhelmed by the development of the fruit. The stamens are reduced to three staminodes that look like the filament of fully developed stamens in the male flower. The floral disc is on the apex of the ovary and forms a ring about the base of the style. The stigma is a complex structure (164-167), usually consisting of a horizontal basal disc with erect finger-like projections arising from its margin to form a cupular structure. The degree of connation among the finger-like projections is variable, with entire stigmas occurring very rarely.

Fruits and Seeds

Fruit characters are some of the most useful for the delimitation and understanding of *Cucumis* species, but they are also those most affected by domestication and improvement. The most decisive and easily used couplets in the Key to the Species of *Cucumis* utilize fruit characters. Attempts were made to construct keys using only floral and inflorescence characters, but they were ineffective.

The fruit-stalk is referred to as a pedicel. As the fruit develops, the lower portion of the hypanthium of the female flower develops into the fruit, while the outer surface of the hypanthium and its ornamentation becomes that of the fruit. The stalk supporting the female flower develops into that supporting the fruit. Therefore, the outer surface of the fruit is homologous to the outer surface of the lower portion of the female-flower hypanthium, and the fruit-stalk is homologous to the stalk of the female flower. The shape of the fruit pedicel (176), cylindrical or flaring upwards, is a very useful character and consistent with that of the female flower described above.

Fruit coloration is very difficult to particularize because of the subjectivity of descriptions, variation during development, and general lack of data. Many fruits are bicolored, having longitudinal stripes running from the base to apex (177). The color combinations are almost equal to the number of observers. This variation cannot be analyzed into developmental effects and vocabulary of the observer, so fundamental colors are recognized herein. Many fruits are described as changing colors or becoming of one color as they mature. The observations are so unsystematic that it is difficult to determine what actually happens in many taxa.

A very useful character in the identification of *Cucumis* species is the presence or absence of aculei on the fruit (183). These structures are referred to as aculei because in the few cases in which a full range of developing fruit is available, the structures on the lower part of the female-flower hypanthium, to referred to as aculei, obviously develop into the structures found on the surface of the fruit. Pubescence (184) and aculei are found together only on mature fruit of *C. prophetarum* subsp. *prophetarum*, *C. thulinianus*, and *C. pubituberculatus*. In *C. prophetarum* subsp. *prophetarum* both glabrous, aculeate fruit and pubescent, aculeate fruit are found. The distribution of pubescent, aculeate fruit is sporadic through the geographic range of the subspecies, which suggests that the condition may have arisen more than once. Extensive studies are needed to properly evaluate the character and understand its genetics.

The seeds are very uniform in their shape, coloration, and smooth surface (192-200) (Singh and Dathan, 1874). Seed size can be used to separate some species, but some cultivated forms have extraordinarily large seeds. Only a single collection of *C. melo* subsp. *melo* (*Must 1507*) from Australia had winged seeds. The wing was small and formed only a small portion of each seed. It looked like the seed coat, of which it is probably an extension.

BIOSYSTEMATIC DATA

In the last 50 years, there have been tremendous gains in our knowledge of the biology of *Cucumis*. This has contributed to improving the classification of *Cucumis*, and significantly influenced the classification presented here. The voluminous literature has been summarized below with an emphasis on findings that affected the classification of *Cucumis*. These studies have depended heavily upon living material held in germplasm collections, consequently numerous wild species have not been studied. Many of these studies do not cite vouchers or the accession numbers of the collections from which materials were obtained. Therefore, it is impossible to repeat those studies or to verify the identifications of the materials used. For example, Esquinas-Alcazar (1977) placed accessions of *C. aculeatus* and *C. sagittatus* in different groups. This could be the result of important biological differences between accessions of these species or misidentifications. Fortunately, the results of these studies agree, in general, with each other and the systematic conclusions presented here. Therefore, in my opinion, the cited studies were well done, and they have been taken into consideration when constructing the classification presented here.

Cucurbitacins

The Cucurbitaceae have long been known for their medicinal properties and bitter constituents (Enslin and Rehm, 1957; Rehm et al., 1960; Rehm, 1960; Watt and Breyer-Brandwijk, 1962), but only recently have the bitter principles been isolated and characterized as a unique series of compounds named cucurbitacins. The genus *Cucumis* was separated from all other Cucurbitaceae by the presence of cucurbitacin B; *C. africanus* and *C. myriocarpus* contained cucurbitacin A; *C. sativus* had only cucurbitacin C; and only *C. sagittatus* had cucurbitacin F. No vouchers are cited, so the identity of the species used cannot be verified.

Flavonoids

Brown et al. (1969) surveyed 15 species of *Cucumis* for their flavonoid compounds and found 26 different compounds. Based on similarities in the distribution of flavonoid compounds, they grouped the studied species into two large complexes and five species with distinct patterns unlike any other species. Their first group consisted of *C. africanus*, *C. heptadactylus*, *C. melo*, and *C. myriocarpus*, and their second of *C. anguria*, *C. dipsaceus*, *C. ficifolius*, *C. prophetarum*, *C. pustulatus*, and *C. zeyheri*. *Cucumis hirsutus*, *C. humifructus*, *C. metuliferus*, *C. sagittatus*, and *C. sativus* were each considered to be distinct from all the other species

11

studied. Unfortunately, they did not report which accessions of *Cucumis* were used in the study. They used material being screened for nematode resistance at the U.S. Vegetable Breeding Laboratory, so it is safe to assume that all their accessions were from those maintained by the U.S. Department of Agriculture.

Isozymes

Large scale isozyme studies were first carried out in the genus *Cucumis* by Dane (1976, 1983) and Esquinas-Alcazar (1977). Dane's study was general in nature, using principally germplasm accessions held by the U.S. Department of Agriculture and a few supplied by the USDA Plant Introduction Station, Ames, Iowa, and by T.W. Whitaker. Esquinas-Alcazar's study focused upon *C. melo* and its cultivars using material obtained from the U.S. Department of Agriculture, the Instituto Nacional de Semillas Selectas, Spain, various research workers, and Spanish farmers, and also surveyed available samples of other species, especially in relation to *C. melo*. Both studies used a combination of Nei's (1972) genetic distance for the isozymes and crossability, along with cytology and other types of data when relevant, to segregate the species of *Cucumis* into four groups. The first group, the *cross compatible* species (Dane, 1976, 1983) or *anguria* group (Esquinas-Alcazar, 1977), was composed of *C. aculeatus* (some accessions), *C. africanus*, *C. anguria*, *C. dipsaceus*, *C. ficifolius*, *C. heptadactylus*, *C. myriocarpus*, *C. pustulatus*, and *C. zeyheri*; the second group, the *metuliferus* group (Esquinas-Alcazar, 1977), was composed of *C. aculeatus* (some accessions), *C. metuliferus*, and *C. sagittatus* (some accessions); the third group, the *melo* group (Esquinas-Alcazar, 1977), was composed of *C. melo* and *C. sagittatus*; and, the fourth group, the *sativus* group (Esquinas-Alcazar, 1977), consisted only of *C. sativus*. Members of the first group were compatible among themselves and not with members of the other three groups, while those of the second, third, and fourth groups had internally incompatible members that were also incompatible with members of the other groups.

Puchalski et al. (1978) also conducted a broad isozyme survey of *Cucumis* without mentioning other types of data or the source of the material studied. In general, their findings agree with those of Dane (1976, 1983) and Esquinas-Alcazar (1977). They also examined two additional species, *C. hirsutus* and *C. humifructus*. Their results indicated that *C. dipsaceus* and *C. heptadactylus* differed from all other species in their isozyme composition, which would exclude them from the first group of Dane (1976, 1983) and Esquinas-Alcazar (1977). Dane (1983) commented that this could be the result of variation in the available samples of these two species and that it emphasizes the major weakness of studies based on a small sample. *Cucumis hirsutus* and *C. humifructus* had distinctive isozyme patterns with some alleles in common with *C. melo*.

Perl-Treves et al. (1985) and Staub et al. (1987, 1991) have also carried out general isozyme surveys of *Cucumis*. Perl-Treves et al. (1985) obtained plant materials from the South African Department of Agriculture, U.S. Department of Agriculture, Institute for Horticultural Plant Breeding, Wageningen, and Weizmann Institute of Science, Israel, and Staub et al. (1987, 1991) used only plant materials of the U.S. Department of Agriculture. Their findings agree in general with those of Dane (1976, 1983) and Esquinas-Alcazar (1977) but differ in some details, as did those of Puchalski et al. (1978). Perl-Treves et al. (1985) examined another species, *C. meeusei*, which fitted into the first group next to *C. dipsaceus*, while *C. humifructus* and *C. sagittatus* formed a pair close to *C. melo* well separated from the remainder of the genus. Staub

(1987) had some collections of *C. anguria* var. *anguria* and *C. myriocarpus* grouped together with *C. metuliferus* in a group well separated from the remainder of the genus.

Puchalski and Robinson (1990) conducted a second isozyme survey of *Cucumis* using accessions held by the New York State Agricultural Experiment State, Geneva, New York. They again confirmed the overall pattern discovered by Dane (1976, 1983) and Esquinas-Alcazar (1977) dividing the genus into five groups, and they made some new observations. They found the isozyme patterns of *C. africanus* to be similar to those of *C. zeyheri* and *C. meeusei*, and placed *C. sagittatus* with the cross compatible species.

Deoxyribonucleic Acid

Perl-Treves (1985) and Perl-Treves and Galun (1985) studied the plastome of *Cucumis*. They used plant materials from the South African Department of Agriculture, U.S. Department of Agriculture, Institute for Horticultural Plant Breeding, Wageningen, and Weizmann Institute of Science, Israel, and these same materials were also used in their isozyme study (Perl-Treves et al., 1985) referred to above. Using restriction enzymes and DNA hybridization, they constructed a map of the *C. melo* plastome. Rather than using the changes in fragment lengths to characterize species, they used restriction sites implying comparison of the mutations giving rise to the varying fragment lengths. They considered restriction site data to be more accurate than fragment length data. The overall outline of their results is very similar to those of the isozyme studies, that is a large *first group*, many of whose species can be crossed, and a number of well separated small groups of one or two species which do not cross among themselves or with members of other groups. Their large group consisted of: *C. africanus*, *C. anguria*, *C. dipsaceus*, *C. ficifolius*, *C. heptadactylus*, *C. meeusei*, *C. myriocarpus*, *C. prophetarum*, *C. pustulatus*, and *C. zeyheri*. The four smaller groups are: 1) *C. humifructus* and *C. metuliferus*; 2) *C. sagittatus*; 3) *C. melo*; and, 4) *C. sativus*. The most obvious difference is that *C. humifructus* was grouped with *C. metuliferus* rather than uniquely separated from the rest of the genus or grouped with *C. sagittatus*. Also, *C. sagittatus* appeared close to *C. melo*, but did not form a group with it. The resolution within the large grouping was at maximum three mutations between some species, and other species were identical. Further studies, perhaps with nuclear DNA, are needed to obtain data with a greater resolving power within the large group.

Cytology

There has long been a desire to transfer disease resistances into *C. melo* and *C. sativus* by crossing with wild species known to have such resistances. Consequently, there have been a number of cytological studies made of *Cucumis* (Kozuchov, 1930; Bhaduri and Bose, 1947; Naithani, 1947; Chandola et al., 1965; Shimotsuma, 1965; Deakin et al., 1971; Singh and Roy, 1974, 1975; Dane and Tsuchiya, 1976, 1979; Varekamp et al., 1982; Ramachandran, 1984; Singh and Yadava, 1984; Singh et al. 1984; Nijs and Visser, 1985; Ramachandran and Narayan, 1985; Raamsdonk et al., 1989; Singh, 1990). After repeated examination, most *Cucumis* species have been found to be 2N = 24: *C. africanus*, *C. anguria*, *C. dipsaceus*, *C. ficifolius*, *C. hirsutus*, *C. humifructus*, *C. melo*, *C. metiluferus*, *C. myriocarpus*, *C. prophetarum*, *C.*

pustulatus, *C. sagittatus*, *C. sacleuxii*, and *C. zeyheri* (complete citations accompany each species description below). Among these species, three have also been reported to be polyploid: *C. ficifolius*, 2N = 48 (Dane and Tsuchiya, 1979; Singh et al. 1984; Nijs and Visser, 1985), *C. pustulatus*, 2n = 48 (Ramachandran, 1984; Nijs and Visser, 1985 Ramachandran and Narayan, 1985) or 72 (Dane and Tsuchiya, 1979), and *C. zeyheri*, 2n = 48 (Dane and Tsuchiya, 1976, 1979; Varekamp et al., 1982; Ramachandran, 1984; Nijs and Visser, 1985; Ramachandran and Narayan, 1985). Two cultivars of *C. melo* have been reported to be aneuploids of 2n = 20 and 22 (Chandola et al., 1965), and three species, *C. aculeatus*, *C. heptadactylus* and *C. meeusei*, have so far been reported only as tetraploids of 2n = 48. *Cucumis sativus* is the only species of *Cucumis* reported to have a chromosome count of 2n = 14.

There are two base chromosome numbers in *Cucumis*, x = 7 and x = 12. Both directions of evolution have been proposed, development of x = 12 from x = 7 by fragmentation (Kozuchov, 1930; Whitaker, 1933; Bhaduri and Bose, 1948; Ayyangar, 1967) or of x = 7 from x = 12 by fusion (Trivedi and Roy, 1970), and rejected (Ramachandran, 1984; Ramachandran and Seshadri, 1986). The most reasonable hypothesis is that x = 7 is derived from x = 12 (Raamsdonk et al., 1989; Singh, 1990). The following data support such a conclusion: 1) Commonest base number in Cucurbitaceae and tribe Melothrieae is x = 12 (Trivedi and Roy, 1970; Jeffrey, 1980); 2) all *Cucumis* polyploids have a base number of x = 12, so x = 12 is older than x = 7 based on Stebbins' (1971) polyploid series clock; 3) *C. sativus* has an increased amount of heterochromatin (Ramachandran et al., 1985); and, 4) fixation of some gene-loci was found only in *C. sativus* (Dane, 1976; Esquinas-Alcazar, 1977) suggesting genetic drift or migration and establishment of a small founder population giving rise to x = 7.

Crossability

As early as 1859, Naudin (1862) attempted to make crosses between cucurbits by removing the male flowers and transferring pollen by hand. Many more recent studies have also attempted to make interspecific crosses in *Cucumis* (Batra, 1953; Andrus and Fassuliotis, 1965; Deakin et al., 1971; Chelliah and Sambandam, 1972; Ruitter, 1974; Fassuliotis, 1977; Dane et al., 1980; Kho et al., 1980; Visser and Nijs, 1983; Singh and Yadava, 1984; Nijs and Visser, 1985; Nijs and Custers, 1990; Chatterjee and More, 1991). Raamsdonk et al. (1989) have summarized the data in two crossing polygons presented in Figs. 2 and 3 (Raamsdonk et al., 1989). The species *C. heptadactylus*, *C. humifructus*, *C. melo*, and *C. sativus* have never been successfully crossed with any other species of *Cucumis* to produce a fertile F_1 generation. The following species can be crossed to a limited extent among themselves: *C. africanus*, *C. anguria*, *C. dipsaceus*, *C. ficifolius*, *C. metuliferus*, *C. myriocarpus*, *C. prophetarum*, *C. pustulatus*, and *C. zeyheri*; refer to the above literature for complete listings of all the successful crosses. This cross-compatible group of species corresponds well to the first group defined by Dane (1976, 1983) and Esquinas-Alcazar (1977) based on isozymes and to the large group obtained by Perl-Treves (1985) and Perl-Treves and Galun (1985) based on the plastome; refer to the discussions of isozymes and deoxyribonucleic acid above.

MATERIALS AND METHODS

Herbarium 5,880 specimens of *Cucumis* and related genera of Cucurbitaceae were borrowed from the following institutions (acronyms according to Holmgren et al., 1990): BH, BLAT, BM, BOL, BR, C, CAL, CANB, CM, E, GAT, HUJ, K, KUN, L, LE, LISC, LISU, LIV, LMU, LY, M, MEL, MO, MPU, NSW, PE, PRE, S, SRGH, TI, UPS, W, and the Vegetable Research Institute, Beijing, and were studied at NA, NY, and US. Selected *Cucumis* accessions held by the U.S. Department of Agriculture were grown out in greenhouses at the Beltsville Agricultural Research Center, Beltsville, Maryland. They were grown in 10 inch pots and trained on upright supports following the cultivation techniques used at the Institute for Horticultural Plant Breeding, Wageningen (Nijs, pers. comm.). The plants were observed throughout their life cycle, and herbarium specimens were prepared and deposited at NA. They are cited in the Index to Exsiccatae under either *Kirkbride* or *Kirkbride & Huette*. All label data from each collection were stored in a fixed-length-field database on a microcomputer database.

Only information on labels accompanying *Cucumis* specimens with identifiable collectors were entered in the database; there were 3,719 collections, i.e., records, with 4,708 specimens in the database. Some collections had duplicate specimens deposited in more than one institution resulting in fewer collections than specimens. Collections examined at NY, P, and US were not loaned, and so are not included in the database.

A unique record number was assigned to each collection and lightly written in pencil in the lower right-hand corner of each duplicate specimen (Table 1). This number was used to sort out and eliminate duplicate records of collections with duplicate specimens. The species and infraspecific identifications were assigned by the author, and the taxon numbers are those which appeared in this monograph. The collectors' names and the location data were in numerous languages using various accents. To enter accents, the entire range of the IBM character set, 1 to 255, was used. The following letters with accents and symbols were input using IBM characters 123 to 255: á, number 160; à, number 133; â, number 131; ã, number 132; ä, number 204; ç, number 135; é, number 130; è, number 138; ê, number 136; í, number 161; ñ, number 164; ó, number 162; ô, number 148; ö, number 206; ü, number 129; and, ±, number 241. If a well-known or prominent location, whether part of the detailed location data or separate, was cited on a label, then it was recorded in the Landmark field. This assisted in determining the position of obscure locations. Elevation above sea level in feet was converted to meters.

Few labels gave latitude and longitude of locations in degrees and minutes. These data were obtained from other sources, including geographical references (Anonymous, 1984; Anonymous, 1985; Gunn and Codd, 1981; Polhill, 1988), country gazetteers of the United States Board on Geographic Names, pencil notes on specimens, and the PRECIS (PRE Computerized Information System) database of the Botanical Research Institute, Pretoria, South Africa (Gibbs Russell, 1985; Morris and Glen, 1978). In PRECIS a grid system is used for determining locations; each degree square is divided into four half-degree squares, 30' by 30', and 16 quarter-degree squares, 15' by 15', indicated by lower-case letters. Therefore, PRECIS data were used in conjunction with other sources, or when only PRECIS data were available, the latitude and longitude were recorded in degrees with zero minutes. In a few cases, place names had been used more than once in the same country, and so it was impossible to determine which of the different sets of coordinates applied to a location.

15

Table 1. Fields in the database of collections studied in-depth.

Field name	Type[1]	Length	Columns	Description
record_numb	N	5	1-5	Unique number of each collection in database
genus	C	7	6-12	Always *Cucumis*
species	C	13	13-25	Species name
rank	C	6	26-31	Rank of the infrapsecific name
infraspecific	C	32	32-63	Infraspecific name
taxon_numb	C	8	64-71	Taxon number in this monograph
collector	C	61	72-132	Collector's initials and full last name
collnum	C	17	133-149	Collector's number
month	N	2	150-151	Number from 1 to 12 corresponding to month
day	N	2	152-153	Number from 1 to 31
year	N	4	154-157	Number from 1752 to 1988
source	C	26	158-183	Botanical institution(s) where deposited[2]
country	C	29	184-212	Follows United States Government (Public Building Office)
state	C	27	213-239	Name appearing on labels
landmark	C	48	240-287	Prominent location on label
location1	C	61	288-348	See footnote [3]
location2	C	61	349-409	See footnote [3]
location3	C	61	410-470	See footnote [3]
location4	C	37	471-507	See footnote [3]
elevat_me	C	9	508-516	Elevation above sea level in meters
latitdeg	N	2	517-518	Degrees of latitude
latitmin	N	2	519-520	Minutes of latitude
latitdir	C	1	521	Latitude direction, N = north; S = south
longideg	N	3	522-524	Degrees of longitude
longimin	N	2	525-526	Minutes of longitude
longidir	C	1	527	Longitude direction, E = east; W = west

[1] C = character, i.e., any character in the IBM character set; N = numeric.
[2] Acronyms according to Holmgren et al. (1990).
[3] Depending on its length, the location may be divided between location fields 1 to 4.

The MS-DOS diskette accompanying this monograph contains the self-uncompressing file CUCUDIST.EXE. The *Cucumis* collections database is compressed within this file. It is an ASCII text file 1,966,435 bytes in length. To uncompress the database, create a suitably named directory on a hard disk, and change to it, making it the default directory. Place the diskette accompanying this monograph in drive A:, and type A:\CUCUDIST at the DOS prompt, and press enter. The file CUCUMIS.DST will be written to the default directory. Each line in CUCUMIS.DST contains the data for one collection of *Cucumis*. Using suitable database software, establish a database structure with the format given in Table 1; the data for each field occurs in the columns given in Table 1. Import the data into the database, and they are ready for use.

The *Cucumis* collections database can be used in numerous ways. Sorting it by collector and collector's number, the collections critically studied and identified can be determined, and the identity of particular collections can be confirmed. Sorted by taxon, i.e., species and infraspecific epithet, the distribution of taxa can be ascertained in the degree of detail desired. Sorted by country and state, the occurrence of taxa in a geographic area can be outlined. Sorting by location is less useful because the location descriptions are so varied that they will not group together. These data can illuminate many aspects of *Cucumis* when different sorts and selections are used.

Cucumis melo and *C. sativus* have 522 and 70 synonyms, respectively. To facilitate the use of the taxonomic text and the search for names, the synonyms of *C. melo* are listed in Appendix 1, and the synonyms of *C. sativus* in Appendix 2.

Morphological, cytological, and macro-distributional data were stored in a microcomputer in DELTA format (Dallwitz, 1974; Dallwitz and Paine, 1986; Dallwitz, 1991; Partridge et al., 1991). The DELTA system for microcomputers (Dallwitz and Paine, 1986; Partridge et al., 1991) was used to produce the keys to and descriptions of taxa. It was also used to produce an online database for microcomputers to run queries concerning the taxa of *Cucumis* or to identify specimens of *Cucumis*.

The format of the keys and descriptions was controlled by the DELTA system. The DELTA character list used for scoring the characteristics of the taxa is given in Appendix 3. Plane shapes are described in the terminology of the Systematic Association Committee for Descriptive Terminology (1962), and hair types follow Lawrence (1955). Jackson (1928), Stearn (1966), and Radford et al. (1974) were consulted for further clarification of some terms. In the keys and descriptions, each character, with all of its states and relevant comments, is separated by either a semicolon or a period from the next character. Comments within parentheses follow the character state to which they apply.

The MS-DOS diskette accompanying this monograph contains two self-uncompressing files, CUCUMIS.EXE and INTKEY30.EXE. The *Cucumis* database is contained in CUCUMIS.EXE, and the program INTKEY, version 3.0, is contained in INTKEY30.EXE. The expanded database is 180 kilo bytes in size, and the expanded software is 350 kilo bytes. They can be installed either on a hard disk or dual floppy diskettes of an IBM compatible microcomputer. For hard disk installation, follow the instructions in the next paragraph and skip the following one, and for floppy diskette installation, skip the next paragraph and go to the following one.

To install the *Cucumis* database and INTKEY on a hard disk, create an appropriately named directory on the hard disk, and change to the newly created directory, making it the default directory. Place the diskette accompanying this monograph in drive A, type A:\INTKEY30 at the DOS prompt, and press enter. The file will decompress itself and install INTKEY in the default directory. Repeat the preceding process with A:\CUCUMIS. If the diskette must be used in drive B: instead, substitute B: for A:. To start the program and access the *Cucumis* database, type INTKEY at the DOS prompt and press enter.

To install the *Cucumis* database and INTKEY on floppy diskettes, place a formated, empty diskette in drive A:, and change to drive A:, making it the default drive. Place the diskette accompanying this monograph in drive B. Type B:\INTKEY30 at the DOS prompt, and press enter. The file will decompress itself and install INTKEY on the default drive. Repeat the preceding process with B:\CUCUMIS. To start the program, place the diskette containing the *Cucumis* database and INTKEY program in drive A:, change to drive A:, and type INTKEY at the DOS prompt and press enter.

As INTKEY starts, it will scroll through its name, version, author, and information on the *Cucumis* database. Highlighted bars at the top and bottom of the screen mark off the INTKEY window in which commands and material generated by the program are displayed. Commands may be typed on the command line directly below the bottom highlighted bar, or they can be selected from the pull-down menus. The bottom highlighted bar is the prompt line giving a brief summary of what can be typed at the current stage. The highlighted bar at the top of the screen is the function line and indicates some of the keys available at the current point in the program to carry out various functions. Single keys are indicated at the right-hand end of the function line, and keys used in combination with the ALT key at the left-hand end of the line. An arrow pointing down on the function line indicates that the down-arrow or PAGE-DOWN keys can be used to scroll through material in the window, and an arrow pointing up means that the up-arrow or PAGE-UP keys can be used. The ALT+M key combination is used to display the pull-down commands menu, and ALT+H displays helpful information about the current activity, function, or command. The ESC key can be pressed at any time to discontinue scrolling through material in the window, to close a pull-down menu, or to abort the current command. To exit from INTKEY, select QUIT from the commands menu, or type 'quit' on the command line and press enter.

When the INTKEY program is started, its defaults are set for the identification of specimens. Examine the specimen to be identified, and determine what its characteristics are. If you are unsure what characters are available in the database, call the commands menu (press the ALT+M key combination), use the cursor arrows to place the highlighted area on the command CHARACTERS, and press the enter key twice. A complete list of the characters will appear in the window to be scrolled through. If a particular structure is present, such as a fruit, then call the commands menu and use FIND. Select CHARACTERS, then type the letters "fruit" on the command line and press enter. All characters in which the letters "fruit" appear will be listed. A third method is to call the commands menu and use BEST. This will list the characters in the order of their separating powers; those with larger values are more useful in separating the remaining taxa. After selecting a character, type its number and press enter. For a numeric character, a window will appear describing it. The value or range of values should be typed on the command line and enter pressed. For multistate characters, a pull-down menu will appear with its states. If the specimen has a single state for the character, move the highlighted area to that state and press enter. If the specimen has more than one state for the character, move the highlighted area to each state of the specimen one-by-one and press space for each state, then press enter. Only those taxa possessing the selected character states will be considered as additional characters and their states are selected. Finally a single taxon will remain. Call the commands menu, and use DESCRIBE to generate a complete description of the remaining taxon to verify the identification of the specimen. To do another identification, call the commands menu and use RESTART to reinitialize INTKEY to do the next identification.

To query the database for information about the taxa, the defaults should be reset. Call the commands menu and use SET. A second pull-down menu of subcommands will appear; select MATCH, and set it to OVERLAP. To discover what characters are available, use the techniques described in the preceding paragraph. For example, to know what taxa occurring in Ethiopia have aculeate fruits, type '230', native distribution in Africa by country, on the command line and press enter, and select Ethiopia from the pull-down menu. Next type '208', presence or absence of aculei on the surface of the fruit, on the command line, and select aculeate on the pull-down menu. The program reports that 11 taxa remain, i.e., there are 11 taxa in Ethiopia with aculeate fruit. To see which taxa remain, type 'taxa rem' on the command line and press enter, and the 11 taxa will be listed in the window. To do another query, call the commands menu and use RESTART to reinitialize INTKEY to do the next query.

TAXONOMY

Jeffrey (1962, 1964, 1967b, 1980a, 1990) has gradually transformed the classification of the Cucurbitaceae from a system based solely on gross morphological characters to one emphasizing nontraditional characters derived from pollen, seed coat anatomy, phytochemistry, and chromosome numbers. Traditionally (Cogniaux, 1881; Cogniaux and Harms, 1924) *Cucumis* has been associated with such genera as *Benincasa* Savi, *Bryonia* Linnaeus, and *Citrullus* Schrader because of its sigmoid anther thecae. Jeffrey (1962) shifted *Cucumis* into an assemblage including *Cucumella* Chiovenda, *Mukia* Arnott, and other genera in tribe Melothrieae Endlicher subtribe Cucumerinae Pax. Subsequent studies on pollen (Marticorena, 1963) and endosperm haustoria (Chopra, 1955; Gulyayev, 1963) have supported this placement. In 1980, Jeffrey (1980a) consolidated tribe Melothrieae into three subtribes and placed *Cucumis* in tribe Melothrieae subtribe Melothriinae, stating that *Cucumis* was most closely related, in descending order, to *Cucumella*, *Oreosyce* Hooker filius, *Myrmecosicyos* C. Jeffrey, *Mukia*, and *Dicaelospermum* C.B. Clarke. In 1990, Jeffrey again reorganized tribe Melothrieae this time with five subtribes and clustered the same five genera with *Cucumis* in the resurrected subtribe Cucumerinae.

Cucumis

Cucumis Linnaeus, Sp. pl. ed. 1, 1011. 1753; Gen. pl. ed 5, 442. 1754.—TYPE: *Cucumis sativus* Linnaeus (lectotype [Britton and Wilson, 1925]).
  Melo Miller, Gard. Dict. abr. ed. 4 [without pagination]. 1754.—TYPE: *Cucumis melo* Linnaeus (lectotype [Swart, 1979]).
  Plants: Herbs, vines, subshrubs; perennial or annual; with or without a woody rootstock; lacking tubers or rarely tuberous (*C. kalahariensis*); monoecious, or rarely dioecious or andromonoecious. Stems procumbent, climbing, or rarely erect; sulcate; not aculeate or rarely aculeate (*C. aculeatus* and *ficifolius*); variously pubescent or rarely glabrous; with nonbreakaway hairs or rarely breakaway hairs. Nodes not geniculate or geniculate. *Leaves*: Simple; petiolate. Petioles not aculeate or rarely aculeate; variously pubescent or rarely glabrous; pubescence a single type on each petiole, or rarely 2 different types uniformly intermixed or 3 different types in distinct zones on each petiole; with nonbreakaway hairs or rarely breakaway hairs. Leaf blades 3- or 5-palmately lobed, trilobate, pentalobate, heptalobate, or entire; with the margin serrate or entire; narrowly to shallowly ovate, or rarely elliptic or triangular in outline; cordate to subcordate, sometimes acute to broadly so or truncate to subtruncate, or rarely obtuse, sagittate, or hastate at the base; with or sometimes without a basal sinus; acute, acuminate, or rarely obtuse at the apex; variously pubescent, or rarely glabrous or aculeate. Central leaf-blade lobe symmetrical; entire or sometimes pinnatifid; narrowly to broadly elliptic, narrowly to very shallowly ovate, sometimes narrowly linear to narrowly oblong, or rarely narrowly to shallowly triangular in outline; narrowly to broadly acute, obtuse,

or rarely acuminate to broadly so the apex. Lateral leaf-blade lobes asymmetrical or sometimes symmetrical; entire or sometimes pinnatifid; narrowly to transversely elliptic, ovate to very shallowly so, linear to broadly oblong, broadly to shallowly triangular, or rarely square in outline; narrowly to broadly acute, obtuse, or rarely broadly acuminate at the apex at the apex. *Tendrils*: Present or rarely absent (*C. rigidus*); solitary or 5-8 at a node in an axillary fascicle (*C. humifructus*); simple or rarely both simple and bifid on individual plants (*C. insignis*); not aculeate or rarely aculeate; variously pubescent or rarely glabrous. *Inflorescences*: Unisexual, or rarely androgynous or gynecandrous. *Bisexual inflorescences*: Racemose; 8-14-flowered; sessile. *Male inflorescences*: A solitary flower, fasciculate, racemose, sometimes paniculate, or rarely modified compound dichasial; 1-18-flowered; sessile or rarely pedunculate. *Male flowers*: Pedicellate; 5-merous. Pedicel terete or rarely sulcate in cross section; variously pubescent or rarely glabrous; without bracteoles or rarely subtended by a bracteole (*C. heptadactylus*). Bracteole subtending pedicel very narrowly elliptic in outline; narrowly long cuneate at the base; acute to narrowly so at the apex; antrorse-strigose at apex; glabrous at base. Hypanthium infundibular or campanulate; variously pubescent or glabrate. Calyx lobes 5 or rarely 4; linear to oblong or narrowly to broadly triangular in outline; acute to narrowly so at the apex; variously pubescent or rarely glabrous. Corolla yellow; infundibular or rarely campanulate; variously pubescent or rarely glabrous outside; glabrous or puberulent inside. Corolla fused into a basal tube. Corolla lobes elliptic to broadly so, ovate to shallowly so, obovate to narrowly so, or rarely oblong or broadly triangular in outline; narrowly to broadly acute or obtuse, and sometimes also mucronate at the apex; variously pubescent or sometimes glabrous. Stamens free; 3; separating from the free portion of the hypanthium above the ovary; two 2-thecate and one 1-thecate. Filaments terete or radially compressed in cross section; glabrous or with basal 1/2-2/3s puberulent and glabrous apically. Anther thecae sigmoid; glabrous with the edges shortly pubescent. Anther connective extended; obovate, oblong to narrowly so, transversely broadly oblong, or ovate; unilobate or rarely bilobate; obtuse or rarely acute at apex; minutely papillate, sometimes smooth, or rarely puberulous, fimbriate or crenulate at the apex; glabrous. Disc cylindrical or rarely consisting of 3 papillae; glabrous. *Female inflorescences*: A solitary flower or rarely fasciclate; 1(-3)-flowered; sessile. *Female flowers*: Pedicellate; 5-merous. Pedicel terete or sulcate in outline; variously pubescent; with nonbreakaway hairs or rarely breakaway hairs; cylindrical or sometimes flaring upwards from a narrower base to a wider apex. Hypanthium hour-glass shaped. The constricted portion and the lower bulge fused to the ovary. Fused portion of hypanthium ellipsoid, cylindrical, or ovoid; aculeate or not so; glabrous or sometimes variously pubescent. Aculei soft or sometimes stiff; with hyaline bristle at apex. The upper bulge of hypanthium free from the ovary. Free portion of hypanthium campanulate; not aculeate or rarely so outside; variously pubescent or rarely glabrous outside; glabrous or puberulent inside. Ovary with 3-5 placentas; with numerous horizontal ovules. Calyx lobes 5, occasionally 4 or 6; linear to narrowly oblong, triangular to narrowly so, or rarely narrowly elliptic or narrowly obovate in outline; narrowly acute to rarely acute at the apex; variously pubescent. Corolla yellow; infundibular; variously pubescent outside; puberulent or glabrous inside. Corolla tube present or occasionally absent. Corolla lobes elliptic to broadly so, ovate to broadly so, obovate to broadly so, or rarely oblong in outline; obtuse and mucronate, or acute to broadly so at the apex. Staminodes present or rarely absent; 3; separating from the free portion of the hypanthium above the ovary; papilliform; glabrous or rarely puberulent. Style terete in cross section; glabrous; subtended by a circular disc or rarely lacking one. Disc glabrous. Stigma cupular; lobate, or sometimes entire or sublobate; with 1-6 or rarely 9 finger-like projections on the margin. Finger-like stigmatic projections papillate or smooth. *Fruit*: A pepo; not geocarpic, maturing above ground and readily visible, or rarely geocarpic, maturing below ground and not visible. Pedicel sulcate or sometimes terete in cross section; variously pubescent or rarely glabrous; with nonbreakaway hairs or rarely breakaway hairs; cylindrical or sometimes flaring upwards from a narrower base to a wider apex. Fruit monocolored or bicolored with longitudinal stripes from base to apex; variously colored, green, white, red, yellow, brown, or orange; with variously colored longitudinal stripes, light or dark green, white, yellow, brown, purple, or orange; ellipsoid, globose, cylindrical, ovoid, obovoid, or rarely spindle-shaped; aculeate or sometimes not so; glabrous or rarely variously pubescent; blunt or rarely rostrate at the apex. Aculei terete or rarely laterally compressed near the base; with hyaline bristle lost or sometimes persistent. *Seeds*: Many; yellowish white; compressed; immarginate; with the edge acute; elliptic or sometimes ovate; smooth; unwinged or rarely apically winged. Wing transversely oblong; truncate at apex. *Chromosome number*: $2n = 14$ or 24, or sometimes 20, 22, 48, or 72.

Distribution: Native in Africa, southwest Asia, Asia, and Australia, and cultivated throughout the world.

Swart (1979) in *Index nominum genericorum* cited *Cucumis melo* as the type of the genus *Melo* and gave Adanson (1763, 2: 138) as the reference. Adanson (1763, 2: 138) did not directly cite a type species, nor did he include the term "type" or its equivalent, article 8.3 of the *International Code Botanical Nomencalture, Berlin, 1988* [ICBN] (Greuter, et al., 1988). Therefore Adanson did not lectotypify the genus *Melo*, and its lectotypification must be attributed to Swart (1979).

Infrageneric Classification of *Cucumis*

Jeffrey (1980) laid the foundations for a modern infrageneric classification when he transferred the genus *Melo* to the rank of subgenus under *Cucumis* and thereby also automatically created the autonym *Cucumis* subgen. *Cucumis*. He proposed four informal species groups under *Cucumis* subgen. *Melo*: 'Metuliferus' group, one species, *C. metuliferus*; 'anguria' group, about 20 species, including *C. anguria*, *C. dipsaceus*, *C. prophetarum*, *C. myriocarpus*, and *C. sacleuxii*; 'melo' group, three species, *C. melo*, *C. sagittatus*, and *C. humifructus*; and 'hirsutus' group, one species, *C. hirsutus*. These four groupings corresponded very closely to those of Dane (1976, 1983) and Esquinas-Alcazar (1977) modified by the findings of others (Deakin et al., 1971; Dane and Tsuchya, 1976; Kroon et al, 1979). The subgenera represented the different base chromosome numbers within the genus, *Cucumis* subgen. *Cucumis* with $x = 7$ and *Cucumis* subgen. *Melo* with $x = 12$, and the informal groups were cross-incompatible.

Raamsdonk et al. (1989) summarized the available biosystematic data for *Cucumis*. They agreed with Jeffrey's (1980) subdivision of the genus into two subgenera founded on base chromosome number, and discussed three informal species groups within *Cucumis* subgen. *Melo* incorporating data from many disciplines, cucurbitacin and flavonoid chemistry, isozymes, DNA, cytology, and crossability. I agree with their analysis and have extended it to include all the species of *Cucumis*, and here propose two new sections and five new series for the species of *Cucumis* subgen. *Melo*. My rank of series corresponds closely to the hierarchical level of the species groups of both Jeffrey (1980) and Raamsdonk et al. (1989), and they are defined principally by biosystematic data supported by morphological characters. My sections are based solely on the presence or absence of aculei on the female hypanthium and fruit, and correspond to the primary divisions of Naudin (1859) and Cogniaux (1881). The series are significant evolutionary groupings, but the sections are convenient morphological groupings that may or may not have evolutionary significance.

My infrageneric classification is summarized in the following synoptic key. Many species, especially the new ones described here and those described by Thulin (1991), have not been adequately investigated biosystematically, and so have been assigned to subgenera, sections, and series on the basis of one or two biosystematic characters or, in many cases, of morphological similarity.

Key to the Subgenera, Sections, and Series of *Cucumis*

1(0). $2n = 14$; mature plants with cucurbitacin C and flavonoid number 1 (Brown et al., 1969); central leaf-blade lobe narrowly triangular, broadly triangular, very broadly triangular, or shallowly triangular in outline; species 31 and 32

Cucumis subgen. *Cucumis*

   $2n = 20\text{-}72$; mature plants with cucurbitacin B, without flavonoid number 1; central leaf-blade lobe narrowly elliptic, elliptic, broadly elliptic, narrowly linear, narrowly oblong, narrowly ovate, ovate, broadly ovate, very broadly ovate, shallowly ovate, or very shallowly ovate in outline; species 1-30 (*Cucumis* subgen. *Melo*)   2

2(1). Fruit not aculeate; fused portion of female hypanthium not aculeate; species 27-30 (*Cucumis* sect. *Melo*)   3

   Fruit aculeate; fused portion of female hypanthium aculeate; species 1-26 (*Cucumis* sect. *Aculeatosi*)   5

3(2). Tendrils solitary; inflorescences unisexual; fruit not geocarpic, maturing above ground and readily visible; style subtended by a circular disc; plants with a woody rootstock   4

   Tendrils 5-8 at a node in an axillary fascicle; inflorescences androgynous; fruit geocarpic, maturing below ground and not visible; style not subtended by a disc; plants without a woody rootstock; species 28   *Cucumis* ser. *Humifructuosi*

4(3). Plants monoecious, or andromonoecious; male inflorescences sessile; filaments glabrous; species 29 and 30   *Cucumis* ser. *Melo*

   Plants dioecious; male inflorescences pedunculate; filaments with basal 1/2-2/3s puberulent and glabrous apically; species 27   *Cucumis* ser. *Hirsuti*

5(2). Mature plants with cucurbitacin D and flavonoids 11 and 15, or 19 (Brown et al., 1969); seeds elliptic   6

   Mature plants without cucurbitacin D or flavonoids 11, 15, or 19; seeds ovate; species 25 and 26   *Cucumis* ser. *Metuliferi*

6(5) Mature plants with cucurbitacin A (except *C. heptadactylus*), with allele 5 of peroxidases-4 (Esquinas-Alcazar, 1977), without chloroplast DNA restriction number 20 of restriction endonuclease PvuII (Perl-Treves and Galun, 1985); species 6-24

Cucumis ser. *Angurioidei*

   Mature plants without cucurbitacin A, with allele 4 or alleles 2 and 3 of peroxidases-4, with chloroplast DNA restriction number 20 of restriction endonuclease PvuII (except *C. africanus*); species 1-5   *Cucumis* ser. *Myriocarpi*

Key to the Species of *Cucumis*

1(0). Tendrils present; stems procumbent, or climbing   2
        Tendrils absent; stems erect   23. *C. rigidus*

2(1). Tendrils solitary; fruit not geocarpic, maturing above ground and readily visible   3
        Tendrils 5-8 at a node in an axillary fascicle; fruit geocarpic, maturing below ground
        and not visible   28. *C. humifructus*

3(2). Tendrils simple   4
        Tendrils bifid or simple on individual plants   14. *C. insignis*

4(3). Fruit not aculeate   5
        Fruit aculeate   10

5(4). Seeds 2.1-2.4 mm wide   30. *C. sagittatus*
        Seeds 2.5-13 mm wide   6

6(5). Fruit blunt at apex, ovoid, ellipsoid, obovoid, globose, or cylindrical in shape; tendrils
        hispid to hispidulous, hirsute, antrorsely strigose, or retrorsely strigose   7
        Fruit rostrate at apex, spindle-shaped; tendrils pilose   26. *C. rostratus*

7(6). Plants monoecious, or andromonoecious; leaf blades 0.8-1.3 times longer than wide   8
        Plants dioecious; leaf blades 1.4-4.3 times longer than wide   9

8(7). Leaf-blade lobe elliptic, narrowly oblong, broadly ovate, shallowly ovate, or very
        shallowly ovate in outline; corolla tube of male flower 0.8-2 mm long; corolla
        tube of female flower 0.8-2.8 mm long; 2n = 20, 22, or 24   29. *C. melo*
        Leaf-blade lobe broadly triangular, very broadly triangular, or shallowly triangular
        in outline; corolla tube of male flower 3.4-4.9 mm long; corolla tube of female
        flower 3.5-6.5 mm long; 2n = 14   31. *C. sativus*

9(7). Internodes 2.5-4 cm long; leaves hirsute; male inflorescences modified compound
        dichasial or racemose, pedunculate, 1-11-flowered; corolla lobes of male flower
        2-8.8 mm long; corolla lobes of female flower 9.6-14.4 mm long; seeds 4-5.6
        mm wide   27. *C. hirsutus*
        Internodes 5-7 cm long; leaves pilose; male inflorescences a solitary flower, sessile,
        1-flowered; corolla lobes of male flower 12-23 mm long; corolla lobes of female
        flower ca 16 mm long; seeds 3.5-3.6 mm wide   24. *C. baladensis*

10(4). Stems and petioles not aculeate   11
        Stems and petioles aculeate   36

11(10). Fruit blunt at apex; central leaf-blade lobe narrowly elliptic, elliptic, broadly elliptic,
        narrowly linear, narrowly ovate, ovate, broadly ovate, very broadly ovate, or
        shallowly ovate in outline   12
        Fruit rostrate at apex; central leaf-blade lobe broadly triangular or very broadly
        triangular in outline   32. *C. hystrix*

12(11). Corolla lobes of male flowers 1.9-13 X 1-9 mm; pedicel of male flower pubescent   13
        Corolla lobes of male flowers 16-20 X 13-17 mm; pedicel of male flower glabrate
        8. *C. carolinus*

13(12). Fruit aculei 0.5-15 mm long   14
        Fruit aculei 20-22 mm long   13. *C. prolatior*

14(13). Pedicel of female flower cylindrical; fruit pedicel cylindrical (when known)   15
       Pedicel of female flower flaring upwards from a narrower base to a wider apex; fruit
       pedicel flaring upwards from a narrower base to a wider apex   34
15(14). Plants monoecious   16
       Plants dioecious   33
16(15). Fruit glabrous   17
       Fruit pubescent (hispidulous, retrorse-strigose, or puberulent)   31
17(16). Fruit aculei terete   18
       Fruit aculei laterally compressed near base   30
18(17). Tendrils glabrate, hispidulous, hirsute, setose, scabrous, hispid, or pilose   19
       Tendrils antrorse-strigose, or retrorse-strigose   27
19(18). Seeds 4-7.8 mm long, 2-4 mm wide, and 0.9-1.6 mm thick   20
       Seeds ca 8.5 mm long, ca 5.5 mm wide, and ca 3.7 mm thick   15. *C. globosus*
20(19). Calyx lobes of female flower 1.2-4.2 mm long; corolla lobes of female flower pubescent
       (hispidulous or puberulent) inside; plants with a woody rootstock   21
       Calyx lobes of female flower 4.8-11 mm long; corolla lobes of female flower
       glabrous inside; plants without a woody rootstock   26
21(20). Seeds elliptic, 2-3.1 mm wide; aculei on hypanthium of female flower 0.6-2.5 mm long,
       with hyaline bristle 0.4-1.2 mm long; corolla lobes of female flower 2-4 mm
       wide   22
       Seeds ovate, 3.3-4 mm wide; aculei on hypanthium of female flower 2.6-2.8 mm
       long, with hyaline bristle ca 1.6 mm long; corolla lobes of female flower 5.1-12
       mm wide   25. *C. metuliferus*
22(21). Stems and petioles hispidulous, pilose, or scabrous; central leaf-blade lobe entire; aculei
       on hypanthium of female flowers 0.6-1.4 mm long; aculei on fruit 0.5-1.2 mm
       long   23
       Stems and petioles antrorse-strigose, or retrorse-strigose; central blade-lobe
       pinnatifid; aculei on hypanthium of female flowers 1.5-2.4 mm long; aculei on
       fruit 1.3-13 mm long   25
23(22). Stems, petioles, leaf blades, and pedicels of female flowers and fruit with nonbreakaway
       hairs; corolla lobes of male flowers 1-3.2 mm wide; plants perennial   24
       Stems, petioles, leaf blades, and pedicels of female flowers and fruit with breakaway
       hairs; corolla lobes of male flowers 4.8-6 mm wide; plants annua   17. *C. sacleuxii*
24(23). Leaf-blade lobes narrowly elliptic, elliptic, broadly elliptic, or transversely elliptic in
       outline; leaf blades hispidulous   10. *C. prophetarum*
       Leaf-blade lobes ovate or broadly ovate in outline; leaf blades scabrous
       22. *C. hastatus*
25(22). Corolla of male flowers glabrous inside; corolla lobes of male flowers obtuse at apex;
       disc of male flowers 1-1.2 mm long and 1.4-1.8 mm in diam; corolla of female
       flowers glabrous inside; seeds 4.3-6 mm long, 2-2.6 mm wide, and 1-1.3 mm
       thick   10. *C. prophetarum*
       Corolla of male flowers puberulent inside; corolla lobes of male flowers acute or
       broadly acute and mucronate at apex; disc of male flowers 0.6-0.9 mm long and
       0.8-1.3 mm in diam; corolla of female flowers puberulent inside; seeds 5.6-6.7
       mm long, 2.8-3.1 mm wide, and 1.4-1.6 mm thick   12. *C. zeyheri*
26(20). Corolla lobes of male flower obovate in outline, acute at apex, 5.5-8.5 X 3-5 mm;
       pedicel of female flower 5-15 mm long; corolla lobes of female flower obovate in
       outline, acute at apex, 6.5-15 X 3-8.5 mm; fruit aculei 4-6.4 mm long; seeds 4-5
       mm long, ca 2 mm wide, ca 1 mm thick   9. *C. dipsaceus*

Corolla lobes of male flower elliptic in outline, obtuse and mucronate at apex, 11-13 X 8.5-9 mm; pedicel of female flower 55-65 mm long; corolla lobes of female flower broadly elliptic in outline, obtuse at the apex, 29-31 X 22-28 mm; fruit aculei 8.5-13 mm long; seeds 6-6.5 mm long, 2.5-3 mm wide, 1.5-1.6 mm thick
3. *C. quintanilhae*

27(18). Petiole pubescence a single type on each petiole; corolla of female flower glabrous inside   28

Petiole pubescence 3 different types in distinct zones on each petiole; corolla of female flower puberulent inside   1. *C. myriocarpus*

28(27). Corolla lobes of male flower 2.4-10 X 1-3.2 mm; calyx lobes of male flowers 0.4-2 mm long; pedicel of female flower 0.5-40 mm long; calyx lobes of female flower 1.2-4.2 mm long; corolla lobes of female flower 2.4-7.6 X 2-5.5 mm; plants perennial   29

Corolla lobes of male flowers 11-13 X 8.5-9 mm; calyx lobes of male flowers 2.6-4.8 mm long; pedicel of female flower 55-65 mm long; calyx lobes of female flower 6.4-7.2 mm long; corolla lobes of female flower 29-31 X 22-28 mm; plants annual   3. *C. quintanilhae*

29(28). Leaf blades hispidulous; corolla of male flowers glabrous inside; corolla lobes of male flowers 2.4-5.6 mm long   10. *C. prophetarum*

Leaf blades antrorse-strigose; corolla of male flowers puberulent inside; corolla lobes of male flowers ca 10 mm long   21. *C. jeffreyanus*

30(17). Stems and petioles scabrous; petioles 2-8.5 cm long; male inflorescences racemose, 5-10-flowered; corolla lobes of male flowers 1.9-5.6 mm long; pedicel of female flower 12-34 mm long; fruit ellipsoid or cylindrical; seeds 4.2-4.8 mm long, 2.2-2.8 mm wide, and 0.9-1.2 mm thick   2. *C. africanus*

Stems and petioles retrorse-strigose; petioles 0.5-1.5 cm long; male inflorescences a solitary flower; corolla lobes of male flowers ca 8 mm long; pedicel of female flower 4.8-5.6 mm long; fruit globose; seeds ca 8.5 mm long, ca 5.5 mm wide, and ca 3.7 mm thick   15. *C. globosus*

31(16). Corolla lobes 2.4-5.6 mm long, hispidulous outside; seeds 4.3-6 mm long; male-flower hypanthium 5-6.5 mm long, hispidulous; male-flower corolla hispidulous outside
32

Corolla lobes 8-12 mm long, scabrous outside; seeds ca 3.5 mm long; male-flower hypanthium 2.8-4.4 mm long, scabrous; male-flower corolla scabrous outside
11. *C. pubituberculatus*

32(31). Leaf blades obtuse, truncate, or cordate at the base, hispidulous on the upper surface; petioles pubescence a single type on each petiole; leaf blades male-flower corolla glabrous inside; fruit bicolored with longitudinal stripes from base to apex, green with light green or white longitudinal stripes; fruit aculei 0.6-2 mm in diam
10. *C. prophetarum*

Leaf blades sagittate at the base, antrorse-strigose on the upper surface; petioles pubescence 2 different types uniformly intermixed on each petiole; male-flower corolla puberulent inside; fruit monocolored, yellow; fruit aculei 3-5 mm in diam
16. *C. thulinianus*

33(15). Plants without a woody rootstock, lacking tubers; nodes geniculate; leaf blades truncate at base, antrorse-strigose on upper surface; leaf-blade lobes 0.1-0.5 cm wide; male inflorescences racemose; pedicel of male flowers subtended by a bracteole; calyx lobes of male Flowers 1.6-2 mm long; corolla lobes of male flowers 2.2-3

X 2.5-2.8 mm; calyx lobes of female flower 1.7-2.2 mm long; corolla lobes of
   female flower 1.2-1.4 X 0.7-1 mm   4. *C. heptadactylus*
Plants with a woody rootstock, tuberous; nodes not geniculate; leaf blades broadly
   acute or subcordate at base, pilose on upper surface; leaf-blade lobes 0.5-3 cm
   wide; male inflorescences fasciculate or a solitary flower; pedicel of male flowers
   without bracteoles; calyx lobes of male flowers ca 7.2 mm long; corolla lobes of
   male flowers 9.6-11.2 X 6.4-8 mm; calyx lobes of female flower 5.9-6.4 mm
   long; corolla lobes of female flower 9.6-11.2 X 4.6-5.4 mm   5. *C. kalahariensis*
34(14). Plants annual, without a woody rootstock; male inflorescences racemose, 3-10-flowered;
   corolla lobes of male flowers broadly acute at apex; pedicel of female flower
   1.5-7 mm long; corolla lobes of female flower obovate or broadly obovate in
   outline, acute or broadly acute at apex; seeds 5-6 mm long, 2-2.5 mm wide, and
   ca 1 mm thick   6. *C. anguria*
Plants perennial, with a woody rootstock; male inflorescences fasciculate or a
   solitary flower, 1-2-flowered; corolla lobes of male flowers obtuse and mucronate
   at apex; pedicel of female flower 8-45 mm long; corolla lobes of female flower
   elliptic or broadly elliptic in outline, obtuse and mucronate at the apex; seeds
   5.2-6.7 mm long, 2.7-3.2 mm wide, and 1.1-1.6 mm thick   35
35(34). Leaf blades antrorse-strigose or scabrous; pedicel of female flower 8-13 mm long;
   hypanthium of female flower antrorse-strigose outside; corolla lobes of female
   flower 4-4.8 mm wide; fruit pedicel antrorse- or retrorse-strigose   20. *C. meeusei*
Leaf blades hispidulous; pedicel of female flower 15-45 mm long; hypanthium of
   female flower puberulent or hispidulous outside; corolla lobes of female flower
   5.2-8 mm wide; fruit pedicel hispidulous   19. *C. pustulatus*
36(10). Petioles glabrous; leaf blades not aculeate on veins below; tendrils aculeate and
   glabrous; hyaline bristle at apex of aculeus on hypanthium of female flower
   0.8-2.3 times as long as opaque base; calyx lobes of female flower 1.2-2 mm
   long; fruit pedicel glabrate; fruit 6-8 cm long and 2.5-4 cm in diam
                                                               18. *C. aculeatus*
Petioles hispidulous; leaf blades aculeate on veins below; tendrils not aculeate and
   hispidulous; hyaline bristle at apex of aculeus on hypanthium of female flower
   4.2-9.5 times as long as the opaque base; calyx lobes of female flower 2.4-3.5
   mm long; fruit pedicel hispidulous; fruit 2.5-4 cm long and 2-3.5 cm in diam
                                                               17. *C. ficifolius*

Cucumis subgen. *Melo*

Cucumis subgen. **Melo** (Miller) C. Jeffrey, J. Linn. Soc., Bot. 81: 238. 1980.—BASIONYM:
   Melo Miller. Species 1-30.
   Cucumis subgen. *Melo* was erected because its base chromosome number is x = 12 while
that of subgen. *Cucumis* is x = 7 and its members have never been successfully crossed with
those of subgen. *Melo*. This division is also supported by the occurrence of cucurbitacin and
flavonoid compounds; see Key to the Subgenera, Sections, and Series of *Cucumis* for a summary
of the characters. This subgenus is generally referred to as the 'African' one. Without doubt its
centers of diversity are in southern and eastern Africa, but its natural range extends from the
Cape of Good Hope northwards through Africa to the southern limits of the Sahara in the west

and reaches the Mediterranean Sea in the east. *Cucumis melo* and *C. prophetarum* are found across all of southwest Asia reaching Pakistan and India, and *C. melo* is found throughout tropical and subtropical Asia, the Pacific, and also in Australia. *Cucumis metuliferus* and *C. pustulatus* reach only the southwestern corner of the Arabian peninsula, and are possibly naturalized there.

Cucumis sect. *Aculeatosi*

Cucumis sect. **Aculeatosi** Kirkbride, sect. nov.—TYPE: *Cucumis dipsaceus* Ehrenberg ex Spach. Species 1-26.

A sect. *Melone* fructibus aculeatis differt.

This section assembles within subgen. *Melo* the series with aculeate fruit, ser. *Myriocarpi*, ser. *Angurioidei*, and ser. *Metuliferi*. It is a convenient morphological grouping that corresponds, in its composition, to the second lead of couplet 4 in the Key to the Species of *Cucumis* and to major divisions of Naudin (1859) and Cogniaux (1881).

Cucumis ser. **Myriocarpi** Kirkbride, ser. nov.—TYPE: *Cucumis myriocarpus* Naudin. Species 1-5.

A ser. *Angurio* plantis maturis sine cucurbitacino *A*, cum ser. *Angurii* plantis infertilibus, sect. *Melonis* fructuum crementum non excitantibus, peroxidasi-4 allelomorphi *4* vel peroxidasi-4 allelomorphi *2* allelomorphique *3* (Esquinas-Alcazar, 1977), chloroplasti acidi deoxyribonecleici PvuII-endonucleasi scissi loco scisso *20* (Perl-Treves and Galun, 1985) absimilis.

Members of *Cucumis* ser. *Myriocarpi* were grouped with the members of *Cucumis* ser. *Angurioidei* by Dane (1976, 1983), Esquinas-Alcazar (1977), Perl-Treves and Galun (1985), and Jeffrey (1980). It is a group of cross-compatible species that are cross-incompatible with members of *Cucumis* ser. *Angurioidei* and can not stimulate fruit set in members of *Cucumis* ser. *Melo* (Raamsdonk, 1989). It is also supported by chemical and DNA characters summarized in the Key to the Subgenera, Sections, and Series of *Cucumis*. *Cucumis heptadactylus* is distinct from the others in the series. It does not cross with other members of the series or stimulate fruit set in *Cucumis* ser. *melo*, and its peroxidase-*4* patterns are distinct (Esquinas-Alcazar, 1977; Raamsdonk, 1989), but its peroxidase-*3* and flavonoid patterns are identical to those of the remainder of the series. It is more advanced than other members of the series because it is dioecious and a tetraploid while the rest of the series is diploid. *Cucumis kalahariensis* is included in this series because of its morphological similarity to *C. heptadactylus*. This series is found only in southern Africa.

1. **Cucumis myriocarpus** Naudin, Ann. Sci. Nat. Bot., sér. 4, 11: 22. 1859.

Plants: Herbs; annual; with a woody rootstock; lacking tubers; monoecious. Stems procumbent; sulcate; not aculeate; retrorse-strigose; with nonbreakaway hairs. Stem hairs 0.4-1(-1.8) mm long. Nodes not geniculate. Internodes 2-9(-20) cm long. *Leaves*: Petioles 1.5-6(-13.5) cm long (cultivated specimens sometimes exhibit long petioles); sulcate; not aculeate; pubescence 3 different types in distinct zones on each petiole; basally retrorse-strigose, hirsute in the middle, and apically antrorse-strigose; with nonbreakaway hairs. Leaf blades 5-palmately lobed; with the margin serrate; broadly ovate to very broadly ovate, or ovate (rarely) in outline; cordate to subcordate at the base; with a basal sinus; 0.5-1.5(-3) cm deep (cultivated specimens exhibit the deepest sinuses); narrowly acute to acute to broadly acute, or obtuse at the apex; 2.5-9(-16) X 2-8.5(-13) cm (cultivated specimens exhibit extreme values); 0.8-1.2 times longer than wide; antrorse-strigose (or sparsely so) on the upper surface; not aculeate on the veins below; antrorse-strigose or hispidulous on the veins below; antrorse-strigose or hispidulous on the intervenium below; with nonbreakaway hairs. Central leaf-blade lobe pinnatifid or entire; elliptic to narrowly elliptic in outline; narrowly acute to acute to broadly acute, or obtuse at the apex; 1.7-6.5(-12.5) X 1-4(-7.5) cm (cultivated specimens exhibit extreme values). Lateral leaf-blade lobes asymmetrical; entire; elliptic, or broadly elliptic to transversely elliptic in outline; broadly acute to obtuse at the apex; 0.5-4.5(-5.5) X 0.4-2.5(-4) cm (cultivated specimens exhibit extreme values). *Tendrils*: Present; solitary; simple; 1.5-4.5 cm long; not aculeate; antrorse-strigose. *Inflorescences*: Unisexual. *Male inflorescences*: A solitary flower, fasciculate, or paniculate; 1-6-flowered; sessile. *Male flowers*: Pedicel terete in cross section; 5-12 mm long; hispidulous or antrorse-strigose; without bracteoles. Hypanthium infundibular or campanulate; 3.6-5 mm long; 2.4-3.6 mm in diam; antrorse-strigose or hispidulous (sparsely). Calyx lobes linear to narrowly triangular in outline; narrowly acute at the apex; 1.6-3.6 X 0.3-0.6 mm; sparsely antrorse-strigose, or hispidulous (sparsely). Corolla infundibular; puberulent or hispidulous outside (sparsely); sparsely puberulent or glabrous inside. Corolla tube 0.6-2.4 mm long; 3-6 mm in diam; sparsely hispidulous or puberulent outside. Corolla lobes elliptic to broadly elliptic in outline; obtuse or broadly acute, and mucronate at the apex; 3.2-6 X 2.8-4.4 mm; sparsely hispidulous or puberulent outside; sparsely puberulent or glabrous inside. Stamens separating from the hypanthium 1.6-2.4 mm from the base of the hypanthium. Filaments terete in cross section; ca 0.4 mm long; 0.2-0.4 mm wide; glabrous. Anther thecae 1.5-2.4 mm long; puberulent. Anther connective obovate; unilobate; obtuse at apex; papillate at the apex; 0.8-1.2 X 0.6-0.8 mm. Disc cylindrical; 0.6-1 mm long; 0.8-1.4 mm in diam. *Female inflorescences*: A solitary flower; 1-flowered. *Female flowers*: Pedicel terete to sulcate in outline; (5-)8-30(-60) mm long; hispidulous or retrorse-strigose; with nonbreakaway hairs; cylindrical. The lower 2/3 or 3/4 of hypanthium fused to the ovary. Fused portion of hypanthium ovoid; 4.4-6.4 mm long; 2-3.2 mm in diam; aculeate; glabrous. Aculei soft; 1-2 mm long. Hyaline bristle at apex of aculeus 0.6-1.2 mm long; 1-1.5 times as long as the opaque base. The upper 1/3 or 1/4 of hypanthium free from the ovary. Free portion of hypanthium 1.2-4 mm long; 1.6-2.4 mm in diam; not aculeate outside; hispidulous (to sparsely so) outside; puberulent (to sparsely so) inside. Calyx lobes linear to narrowly triangular in outline; narrowly acute at the apex; 1-4.4 X 0.2-0.4 mm; sparsely hispidulous. Corolla puberulent outside; puberulent (to sparsely so) inside. Corolla tube present; 0.3-1.2 mm long; 2-4 mm in diam; puberulent outside. Corolla lobes elliptic to broadly elliptic in outline; broadly acute, or obtuse and mucronate at the apex; 2-6.8 X 1.8-3.6 mm; puberulent outside; puberulent (to sparsely so) inside. Staminodes present; separating from the free portion of the hypanthium 0.6-1.6 mm above the ovary; 0.6-0.8 mm long; 0.1-0.2 mm in diam; glabrous. Style 1-2 mm long; 0.2-0.6 mm in diam; subtended by a circular disc. Disc 0.3-0.8 mm long; 0.7-1.2 mm in diam. Stigma 1-2.2 mm long; 1.1-1.8 mm in diam; lobate; with 6, 5, or 3 finger-like projections on the margin. Finger-like stigmatic projections 0.6-1.7 mm long; 0.3-0.6 mm in diam; papillate. *Fruit*: Not geocarpic, maturing above ground and readily visible. Pedicel sulcate or terete in cross section; 1-6 cm long; hispidulous or retrorse-strigose; with nonbreakaway hairs; cylindrical. Fruit bicolored with longitudinal stripes from base to apex or monocolored (when mature); green (when immature), orange (when mature), or yellow (when mature); with light green (light, when immature), white (when immature), brown (when immature), purple (when immature), or yellow (when mature) longitudinal stripes; ellipsoid or globose; 1.5-3(-5) cm long; 1.5-2.5 cm in diam; aculeate (to sparsely so, dark longitudinal bands with 1-20 aculei and the light longitudinal bands with

0-5(-10) aculei); glabrous; blunt at the apex. Aculei terete; with hyaline bristle lost; 0.5-0.6 or 1.5-4 mm long; 0.2-0.5 mm in diam. *Seeds*: Seeds elliptic; 4.6-6.7 X 2.4-3 mm; 1.1-1.6 mm thick; unwinged. *Chromosome number*: 2n = 24 (Kozuchov, 1930; Shimotsuma, 1965; Deakin, Bohn & Whitaker, 1971; Dane & Tsuchiya, 1976; Singh & Yadava, 1984; Yadava, Singh & Arya, 1984; Nijs & Visser, 1985).
Distribution: Africa: Lesotho, Mozambique, South Africa, and Zambia.

Key to the Subspecies of *Cucumis myriocarpus*

1. Mature fruit longitudinally striped in various shades of orange with (10-)15-20 aculei on the darker and 2-5(-10) on the lighter stripes of young fruit; aculei 1.5-4 mm long; Botswana, Lesotho, Mozambique, South Africa, and Zambia.
   C. myriocarpus subsp. *myriocarpus*
1. Mature fruit pale yellow without longitudinal stripes or rarely with faint traces of them, with 1-4(-10) aculei on the darker longitudinal stripes and none on the lighter longitudinal stripes of young fruit; aculei 0.5-0.6 or 2.9-3.2 mm long; Lesotho and South Africa (Cape Province, Transvaal, and Natal).
   C. myriocarpus subsp. *leptodermis*

1a. **Cucumis myriocarpus** subsp. **myriocarpus**—TYPES: SOUTH AFRICA. Cape Province: Somerset, *J.H. Bowker s.n.* (lectotype, here designated: K!).—Orange Free State: Vet River, *J. Burke s.n.* (syntype: K!).—*Cucumis africanus* Linnaeus filius var. *myriocarpus* (Naudin) J.B. Davy, Man. pl. Transvaal 1: 39. 1926.
Cucumis dissectifolius Naudin, Ann. Sci. Nat. Bot., sér. 4, 11: 23. 1859.—TYPES: SOUTH AFRICA. Transvaal: Mooye River, *J. Burke 276* (lectotype, K), *488* (syntype: K!).—Cape Province: Grahamstown, *Ward s.n.* (syntype: K!).
Cucumis grossularia Hort. in Voss, Vilm. Blumengärten. ed. 3 1: 344. 1894. Invalid, Voss cited it as a synonym of *C. myriocarpus* Naudin.
Cucumis africanus Linnaeus filius var. *acutilobus* Cogniaux, Bull. Herb. Boissier 3: 418. 1895.—TYPES: SOUTH AFRICA. Transvaal: Houtbosh, *A. Rehmann 6311* (lectotype: K!; isolectotypes: BM!, BR!); *A. Rehmann 5169* (syntype: not seen).
Cucumis merxmuelleri Süssengüth, Trans. Rhodesia Sci. Assoc. 43: 135. 1951.—TYPE: ZIMBABWE. Marandellas, 28 Apr 1942, *G. Dehn 746* (holotype: M!).

Plants: Internodes 5-9(-20) cm long (the basal internode longer than the apical ones). *Leaves*: Leaf blades broadly ovate to very broadly ovate, or ovate (rarely) in outline; cordate to subcordate at the base; narrowly acute to acute to broadly acute, or obtuse at the apex (rarely); 2.5-9(-16) X (2-)3-8.5(-13) cm; 0.8-0.9(-1.2) times longer than wide; not aculeate on the veins below; antrorse-strigose on the veins below; antrorse-strigose on the intervenium below. Central leaf-blade lobe narrowly acute to acute to broadly acute, or obtuse (rarely) at the apex; 1.7-6.5(-12.5) X 1-4(-7.5) cm. Lateral leaf-blade lobes elliptic in outline; broadly acute to obtuse at the apex; 0.5-4.5(-5.5) X 0.4-2.5(-4) cm. *Male inflorescences*: Paniculate or a solitary flower. *Male flowers*: Pedicel 5-12 mm long; hispidulous or antrorse-strigose. Hypanthium 4-5 mm long; antrorse-strigose. Calyx lobes 1.6-3.6 X 0.3-0.6 mm; sparsely antrorse-strigose. Corolla tube 0.6-1.4 mm long. Corolla lobes 3.2-4.8 X 2.8-4 mm. *Female flowers*: Pedicel (5-)10-30(-60) mm long; hispidulous or retrorse-strigose. Aculei 1-1.8 mm long. Hyaline bristle at apex of aculeus 0.6-1 mm long; 1.3-1.5 times as long as the opaque base. Free portion of hypanthium 2.4-4 mm long. Calyx lobes 1.6-4.4 mm long. Corolla tube 0.6-1.2 mm long. Corolla lobes 4-6.8 X 2-3.6 mm. Stigma 1.8-2.2 mm long; with 6 finger-like projections on the margin. Finger-like stigmatic projections 1.4-1.7 mm long. *Fruit*: Fruit bicolored with longitudinal stripes from base to apex (when mature); green (when immature) or orange (when mature); with light green (when immature), purple (light, when immature), or orange (light, when mature) longitudinal stripes; aculeate (dark longitudinal bands with (10-)15-20 aculei, and the light longitudinal bands with 2-5(-10) aculei); glabrous. Aculei 1.5-4 mm long.

Distribution: Africa: Lesotho, Mozambique, South Africa, and Zambia.

1b. **Cucumis myriocarpus** subsp. **leptodermis** (Schweickerdt) Jeffrey & Halliday, Kew Bull. 30(3): 481. 1975.—*Cucumis leptodermis* Schweickerdt, S. African J. Sci. 30: 460. 1933.—TYPES: SOUTH AFRICA. Cape Province: 11 miles S of De Aar, Mar 1933, *H.G. Schweickerdt 1244* (holotype: PRE!); Zuurvlakte, Aliwal North, Apr, *H.F. Becker 11418* (paratype: PRE!); Middleburg, Apr 1924, *G. Gill 91* (paratype, PRE!); near Victoria West Reservoir, 20 Feb 1926, *C.A. Smith 2405* (paratype: PRE!).—Orange Free State: Fauresmith, alongside road between Phillopolis & Fauresmilk, Mar 1933, *H.G. Schweickerdt 1284* (paratype: W!).

Plants: Internodes 2-6.5 cm long. *Leaves*: Leaf blades broadly ovate in outline; cordate at the base; obtuse at the apex; 2.5-6.5(-12) X 2-5.5(-12) cm; 0.8-1.2 times longer than wide; not aculeate on the veins below; antrorse-strigose or hispidulous on the veins below; antrorse-strigose or hispidulous on the intervenium below. Central leaf-blade lobe obtuse at the apex; 2-3(-5.5) X 1.2-3(-7) cm. Lateral leaf-blade lobes broadly elliptic to transversely elliptic in outline; obtuse at the apex; 0.5-1.5(-3) X 0.5-1.5(-4) cm. *Male inflorescences*: Paniculate, a solitary flower or fasciculate. *Male flowers*: Pedicel 9-12 mm long; hispidulous. Hypanthium 3.6-4.4 mm long; sparsely hispidulous. Calyx lobes 2.4-3.5 X 0.5-0.6 mm; sparsely hispidulous. Corolla tube 1.6-2.4 mm long. Corolla lobes 5.2-6 X 3.6-4.4 mm. *Female flowers*: Pedicel 8-16 mm long; hispidulous. Aculei 1.6-2 mm long. Hyaline bristle at apex of aculeus 0.8-1.2 mm long; 1-1.5 times as long as the opaque base. Free portion of hypanthium 1.2-1.8 mm long. Calyx lobes 1-1.4 mm. Corolla tube 0.3-0.8 mm long. Corolla lobes 2-3.2 X 1.8-2 mm. Stigma 1-1.8 mm long; with 6, 5, or 3 finger-like projections on the margin. Finger-like stigmatic projections 0.6-1 mm long. *Fruit*: Fruit monocolored (when mature, or with faint longitudinal stripes); dark green (when immature) or yellow (when mature); with light green (when immature) or yellow (when mature and stripping weakly evident) longitudinal stripes; sparsely aculeate (dark longitudinal bands with 1-4(-10) aculei, and the light longitudinal bands with none); glabrous. Aculei 0.5-0.6 or 2.9-3.2 mm long.

Distribution: Africa: Lesotho and South Africa (Cape Province, Orange Free State, and Transvaal).

Naudin (1862, 1866) reported the hybridization in cultivation of *C. myriocarpus* with *C. anguria*, *C. dipsaceus*, and *C. figarei*. Later studies (Meeuse, 1962; Deakin et al., 1971; Dane et al., 1980) have confirmed the first two crosses and reported on hybridization of *C. myriocarpus* with *C. africanus*, *C. heptadactylus*, *C. leptodermis*, *C. prophetarum*, and *C. zeyheri*. Of all the crosses, only that with *C. leptodermis* was fully fertile. Jeffrey and Halliday (1975) cited the later studies as evidence for the reduction of *C. leptodermis* to a subspecies under *C. myriocarpus*.

The petioles of *C. myriocarpus* have an unusual pubescence consisting of three different types of hair. The base of the petiole is retrorsely strigose, the middle is hispidulous, and the apex is antrorsely strigose. The transition between each hair type is gradual so that there are three zones of distinct pubescence and two transition zones. No other species of *Cucumis* exhibits this type of petiole pubescence.

Schweickerdt (1933) differentiated *C. leptodermis* from *C. myriocarpus* by mature fruit concolorous and pale greenish-yellow, with few "echinae" [aculei], exocarp thin, membranous, and flexible, and fruits usually spherical. In the Latin description of *C. leptodermis* he numerically described the density of aculei as 1-4 on the dark longitudinal bands and none on the lighter bands of young fruit, whereas *C. myriocarpus* was described as having ±20 aculei on the darker bands and 2-5 on the lighter bands. For the identification of herbarium specimens, the number of aculei is the most consistently available characteristic.

In the protologue of *C. myriocarpus*, Naudin (1859) cited references on cultivated plants passing as *C. prophetarum* in the sense of *C. myriocarpus* and not in the sense of *C. prophetarum* Linnaeus, and to two collections in J.D. Hooker's personal herbarium, *J.H. Bowker s.n.* from Somerset and *J. Burke s.n.* from Vet River, cited above. When Schweickerdt (1933) published *C. leptodermis* as a new species, he also redescribed *C. myriocarpus* and cited specimens. Meeuse (1962) considered Schweickerdt's citation of *J. Burke s.n.* from Vet River to be the "holotype," i.e., a lectotypification. Schweickerdt labeled his collection *1244* as "type" of *C. leptodermis*. Consequently he understood the concept of typification and his citation of the *Burke* collection was not meant as a typification but rather citation of a specimen examined. The *Burke* and *Bowker* collections at K were therefore syntypes. Both collections represent *C. myriocarpus* subsp. *myriocarpus*. The *Burke* collection has only young flower buds, while the *Bowker* collection has male and female flowers at anthesis and fruits in various stages of development up to and including maturity. Therefore, I have designated the *Bowker* collection at K as lectotype.

2. **Cucumis africanus** Linnaeus filius, Suppl. pl. 423. 1782.—TYPE: Hermann, *Parad. bat. Tab. 134.* 1698 (holotype).

   Cucumis hookeri Naudin, Gard. Chron. 46: 1503. 1870.—TYPE (here designated): FRANCE. Jardin d'Expér. de Collioure, de l'Intérieur de l'Afrique, graines envoyées par M. Jos. Hooker, 1871-4, *C. Naudin s.n.* (neotype: P!; isoneotypes: P!).

Plants: Herbs; perennial; with a woody rootstock; lacking tubers; monoecious. Stems procumbent; sulcate; not aculeate; scabrous (hairs with a bulbous base); with nonbreakaway hairs. Stem hairs 0.7-0.9 mm long. Nodes not geniculate. Internodes 3-11(-12.5) cm long. *Leaves*: Petioles 2-7.5(-8.5) cm long; sulcate; not aculeate; pubescence a single type on each petiole; scabrous (to densely so, hairs with a bulbous base); with nonbreakaway hairs. Leaf blades 5-palmately lobed or pentalobate (rarely); with the margin serrate; broadly ovate in outline; cordate at the base; with a basal sinus; (0.4-)0.8-1 cm deep; broadly acute or broadly acuminate at the apex; (1.5-)2.5-11.5 X (1.5-)2-7.5 cm; 1-1.2 times longer than wide; sparsely scabrous on the upper surface; not aculeate on the veins above; scabrous on the veins below (with longer hairs than intervenium pubescence); scabrous (to sparsely or densely so) on the intervenium below; with nonbreakaway hairs. Central leaf-blade lobe entire; elliptic in outline; broadly acute or broadly acuminate at the apex; (0.8-)3.5-4.5(-9) X (0.7-)2-3 cm. Lateral leaf-blade lobes symmetrical or asymmetrical; entire; elliptic in outline; obtuse or broadly acuminate at the apex; (0.3-)1-3(-4.5) X (0.3-)0.7-1.5 cm. *Tendrils*: Present; solitary; simple; 2-4.5 cm long; not aculeate; basally scabrous. *Inflorescences*: Unisexual. *Male inflorescences*: Racemose; 5-10-flowered; sessile. *Male flowers*: Pedicel terete in cross section; 3.2-12.8 mm long; sparsely scabrous; without bracteoles. Hypanthium infundibular; 3.5-6.4 mm long; 2.7-3.6 mm in diam; scabrous. Calyx lobes narrowly triangular in outline; narrowly acute at the apex; 1.2-3 X 0.3-0.9 mm; sparsely scabrous. Corolla infundibular; sparsely hispidulous outside; glabrous inside. Corolla tube 1.4-3.2 mm long; 4-6 mm in diam; sparsely hispidulous outside. Corolla lobes broadly ovate, shallowly ovate, or very broadly ovate in outline; broadly acute at the apex; 1.9-5.6 X 3.5-4 mm; sparsely hispidulous outside; glabrous inside. Stamens separating from the hypanthium ca 2.4 mm from the base of the hypanthium. Filaments radially compressed in cross section; 0.6-0.8 mm long; 0.4-0.8 mm wide; glabrous. Anther thecae 2.2-2.8 mm long; glabrous. Anther connective oblong or narrowly oblong; unilobate; obtuse at apex; papillate at the apex; 0.8-1.2 X 0.2-0.8 mm. Disc cylindrical; 0.8-1.1 mm long; 1.4-2 mm in diam. *Female inflorescences*: A solitary flower. *Female flowers*: Pedicel sulcate in outline; 12-34 mm long; scabrous; with nonbreakaway hairs; cylindrical. The lower 2/3 of hypanthium fused to the ovary. Fused portion of hypanthium ellipsoid or cylindrical; 5.6-14.4 mm long; 3.2-4.8 mm in diam; aculeate; glabrous. Aculei soft; 1.3-2.4 mm long. Hyaline bristle at apex of aculeus 0.4-0.8 mm long; 2-2.2 times as long as the opaque base. The upper 1/3 of hypanthium free from the ovary. Free portion of hypanthium 3.2-4.8 mm long; 2.4-5.3 mm in diam; not aculeate outside; hispidulous outside; glabrous inside. Calyx lobes linear in outline; narrowly acute at the apex; 1.6-4 X 0.3-1.2 mm; hispidulous. Corolla sparsely hispidulous outside; sparsely puberulent or glabrous inside. Corolla tube present; 1.2-2.8 mm long; 3.6-4.4 mm in diam; sparsely hispidulous outside. Corolla lobes broadly elliptic in outline; broadly acute at the apex; 4.8-7.9 X 4-6 mm; hispidulous outside; sparsely hispidulous inside. Staminodes present; separating from the free portion of the hypanthium 1.2-1.6 mm above the ovary; 0.5-1 mm long; 0.2-0.4 mm in diam; glabrous. Style 1.2-2 mm long; 0.8-1.2 mm in diam; subtended by a circular disc. Disc 0.6-1 mm long; 1-2.2 mm in diam. Stigma 2.4-3.6 mm long; 2.4-4.6 mm in diam; lobate; with 6 finger-like projections on the margin. Finger-like stigmatic projections 1.6-2.4 mm long; 0.5-0.7 mm in diam; papillate. *Fruit*: Not geocarpic, maturing above ground and readily visible. Pedicel sulcate in cross section; 1-2.5 cm long; hispidulous or antrorse-strigose; with nonbreakaway hairs; cylindrical. Fruit bicolored with longitudinal stripes from base to apex; greenish white; with purplish brown or yellow longitudinal stripes; ellipsoid, or cylindrical; 3-8 cm long; 2-4 cm in diam; aculeate; glabrous; blunt at the apex. Aculei laterally compressed near the base; with hyaline bristle lost; 1-7 mm long; 0.8-4 mm in diam. *Seeds*: Elliptic; 4.2-4.8 X 2.2-2.8 mm; 0.9-1.2 mm thick; unwinged. *Chromosome number*: 2n = 24 (Deakin, Bohn & Whitaker, 1971; Dane & Tsuchiya, 1976; Ramachandran, 1984; Yadava, Singh & Arya, 1984; Ramachandran & Narayan, 1985).

Distribution: Africa: Angola, Botswana, Namibia, South Africa (Cape Province and Transvaal), and Zimbabwe.

When Linnaeus filius (1767) published *Cucumis africanus*, he presented: 1) The binomial; 2) a nine word phrase-name; 3) a description of stems, leaves, flowers, and fruits; 4) origin of the species, Cape of Good Hope; and 5) a reference to Hermann (1698) citing Hermann's phrase-name, page of occurrence, 133, and figure depicting the plant, *t. 134*. Linnaeus filius associated Hermann's CUCUMIS *Africanus echinatus minor* (Hystrix *vegetabilis vulgo*) with figure 134 labelled CUCUMIS *echinatus Colocynthidis foliio* (*fructu majori*). Hermann stated under CUCUMIS *Africanus echinatus minor*, "visum est minoris tantum iconem proponere, cum major sola partium magnitudine à minori discrepet," i.e., only an illustration of the smaller is presented for viewing, because the larger differs from the smaller only in the size of the parts. Figure 134 is incorrectly labelled, and Linnaeus filius correctly cited it as the figure of CUCUMIS *Africanus echinatus minor*. Therefore, the protologue of *C. africanus* refers to the description on page 133 and figure 134.

Naudin (1859) associated Hermann's CUCUMIS *Africanus echinatus minor*, incorrectly citing it, with *C. africanus*, and identified figure 134 and CUCUMIS *echinatus Colocynthidis foliio fructu majori* as *C. anguria* var. *anguria*. Hermann supplied only a diagnosis for CUCUMIS *echinatus Colocynthidis foliio fructu majori* stating that its fruit are larger than CUCUMIS *Africanus echinatus minor* and that it was found growing among melons in the *Maresian* garden. Without a description and based only on its early appearance in Europe, it might be *C. anguria* var. *anguria*. However, the plant depicted in figure 134 is not *C. anguria* var. *anguria* because its short fruiting pedicels are not flaring upwards from a narrow base to a wider apex, and the aculei on the fruit are longer. Subsequent authors (Harvey and Sonder, 1862; Cogniaux, 1881, 1924) followed Naudin. Meeuse (1962) pointed out Naudin's errors, and cited a typotype in Hermann's herbarium at BM as the type of *C. africanus*, possibly volume 84, page 137 (C.E. Jarvis, pers. comm.; C. Jeffrey, pers. comm.).

Since no authentic specimens examined by Linnaeus filius have been found, Hermann's figure 134 is the only element available and therefore the holotype of *C. africanus*, Article 7.3, Note 1, ICBN, Berlin (Greuter et al., 1988).

There are two fruit types in this species, small ellipsoid and large cylindrical. Cylindrical fruits are encountered throughout the range of the species, while ellipsoid fruits are found only in Angola, Namibia, and South Africa. Meeuse (1962) reported that large cylindrical fruits are nonbitter and used as a water source by native peoples and that small ellipsoid fruit are usually bitter and poisonous. There are three types of bitter fruits with increasing degrees of bitterness.

3. **Cucumis quintanilhae** R. Fernandes & A. Fernandes, Revista de Biologia 3(2-4): 269. 1963.—TYPES: BOTSWANA. Turn off to Limpopo Ranches, 10 miles north of Beitbridge, 25 Mar 1959, *R.B. Drummond 6025* (holotype: SRGH!); *A.W. Exell, F.A. Mendonça, & H. Wild 398* (paratypes: BM! LISC!).

Plants: Herbs; annual (probably); monoecious. Stems procumbent; sulcate; not aculeate; hispidulous or retrorse-strigose; with nonbreakaway hairs. Stem hairs 0.6-1.1 mm long. Nodes not geniculate. Internodes (2.5-)4.5-9.5 cm long. *Leaves*: Petioles (1.5-)2-3.5(-5) cm long; sulcate; not aculeate; pubescence a single type on each petiole; hispidulous to antrorse-strigose, or retrorse-strigose; with nonbreakaway hairs. Leaf blades 5- or 3-palmately lobed, or entire (only *J.C. Scheepers 1120*, Transvaal, South Africa); with the margin serrate (to weakly so); broadly ovate in outline; cordate at the base; with a basal sinus; (0.5-)1-1.5 cm deep; broadly acute or obtuse at the apex; 5-10 X 3-8 cm; 1.2-1.7 times longer than wide; scabrous or antrorse-strigose (only *J.C. Scheepers 1120*, Transvaal, South Africa) on the upper surface; not aculeate on the veins below; hispidulous on the veins below; scabrous on the intervenium below; with nonbreakaway hairs. Central leaf-blade lobe entire; elliptic in outline; broadly acute or obtuse at the apex; 4-5.5 X (1-)2.5-3 cm. Lateral leaf-blade lobes symmetrical; entire; broadly elliptic in outline; obtuse at the apex; 1.5-2.5 X 1-2.5 cm. *Tendrils*: Present; solitary; simple; 2-4(-8) cm long; not aculeate; hispidulous to antrorse-strigose, or retrorse-strigose. *Inflorescences*: Unisexual. *Male inflorescences*: A solitary flower or rarely racemose; sessile. *Male flowers*: Pedicel terete in cross section; 15-70 mm long; hispidulous (with pubescence on the basal ½ much sparser than on the apical ½); without bracteoles. Hypanthium campanulate; 5.2-6 mm long; 3.5-4 mm in diam; hispidulous or antrorse-strigose (only *J.C. Scheepers 1120*, Transvaal, South Africa). Calyx lobes narrowly triangular or linear (only *J.C. Scheepers 1120*, Transvaal, South Africa) in outline; narrowly acute at the apex; 2.6-4.8 X 0.2-1 mm; hispidulous or antrorse-strigose (only *J.C. Scheepers 1120*, Transvaal, South Africa). Corolla infundibular; puberulent outside; glabrous or puberulent inside. Corolla tube 1.2-4 mm long; 8-11 mm in diam; puberulent outside. Corolla lobes elliptic in outline; mucronate and obtuse at the apex; 11-13(-20) X 8.5-9(-10) mm; glabrate outside; glabrous inside. Stamens separating from the hypanthium ca 2.8 mm from the base of the hypanthium. Filaments terete in cross section; ca 0.8 mm long; ca 0.4 mm wide; glabrous. Anther thecae 2.6-2.8 mm long; hispidulous. Anther connective obovate; unilobate; obtuse at apex; papillate at the apex; ca 1.6 X 0.8-1 mm. Disc cylindrical; ca 1.4 mm long; ca 2.2 mm in diam. *Female inflorescences*: A solitary flower. *Female flowers*: Pedicel terete in outline; 55-65 mm long; hispidulous to antrorse-strigose; with nonbreakaway hairs; cylindrical. The lower 2/3 of hypanthium fused to the ovary. Fused portion of hypanthium ellipsoid or ovoid; 9.6-12 mm long; 4-4.8 mm in diam; aculeate; glabrous. Aculei 1.6-2 mm long. Hyaline bristle at apex of aculeus 0.5-0.6 mm long; ca 0.5 times as long as the opaque base. The upper 1/3 of hypanthium free from the ovary. Free portion of hypanthium 4-5.6 mm long; ca 5.6 mm in diam; not aculeate outside; antrorse-strigose outside. Calyx lobes linear or narrowly triangular in outline; narrowly acute at the apex; 6.4-7.2 X 0.6-0.7 mm; antrorse-strigose. Corolla puberulent outside. Corolla tube present; ca 4.4 mm long; ca 5.6 mm in diam; puberulent outside. Corolla lobes broadly elliptic in outline; obtuse at the apex; 29-31 X 22-28 mm; puberulent outside. *Fruit*: Not geocarpic, maturing above ground and readily visible. Pedicel sulcate in cross section; ca 4 cm long; glabrate; with nonbreakaway hairs; cylindrical. Fruit monocolored (fide Halliday & Jeffrey, 1978); green (when immature) or yellow (when mature); ellipsoid; ca 4 cm long; ca 2.5 cm in diam; aculeate; glabrous; blunt at the apex. Aculei terete; with hyaline bristle lost; 8.5-13 mm long; 0.8-1.6 mm in diam. Bristle 0.5-0.6 mm long (when present). *Seeds*: Elliptic; 6-6.5 X 2.5-3 mm; 1.5-1.6 mm thick; unwinged.

Distribution: Africa: southernmost Botswana and South Africa (northernmost Transvaal Province).

This a rare species endemic to the border between Botswana and South Africa near Beitbridge-Messina. Additional flowering and fruiting collections are needed, especially from the region of the type-locality, for a more comprehensive evaluation of the species.

4. **Cucumis heptadactylus** Naudin, Ann. Sci. Nat., ser. 4, 11: 24. 1859.—TYPES: SOUTH AFRICA. Cape Province: Near Colesberg, Vanderswaltfontain, *J. Burke 139* (syntypes: BM! K!); Winterfeld, *J.F. Drège 8183* (syntypes: L! MO! P! S! W!); Colesberg, *C. Zeyher 591* (lectotype: P!; isolectotypes: BM! E! LE! MEL! P! S! W!).—Orange Free State: Caledon River, *J. Burke s.n.* (syntypes: BM! K!), *C. Zeyher 590* (syntypes: BM! E! K! P! S!).

Plants: Herbs; perennial; without a woody rootstock; lacking tubers; dioecious. Stems procumbent; sulcate (or weakly so); not aculeate; shortly hirsute to glabrate; with nonbreakaway hairs. Stem hairs 0.5-1 mm long. Nodes geniculate. Internodes 2.5-7 cm long. *Leaves*: Petioles 0.4-2 cm long; sulcate; not aculeate; pubescence a single type on each petiole; shortly hirsute (to sparsely so); with nonbreakaway hairs. Leaf blades 5-palmately lobed (or sometimes 6-8 lobed); with the margin entire; ovate in outline; truncate at the base; without a basal sinus; narrowly acute at the apex; 4.5-9(-15) X 2.5-6 cm; 1.5-1.8(-2.5) times longer than wide; antrorse-strigose (to densely so) on the upper surface; not aculeate on the veins below; antrorse-strigose to glabrous on the veins below; glabrous on the intervenium below; with nonbreakaway hairs. Central leaf-blade lobe entire; narrowly linear in outline; narrowly acute at the apex; 2.7-5.6(-14.5) X 0.1-5 cm. Lateral leaf-blade lobes asymmetrical; entire; linear in outline; narrowly acute at the apex; 0.7-3.2 X 0.1-0.3 cm. *Tendrils*: Present; solitary; simple; 1-2(-4) cm long; not aculeate; shortly hirsute to glabrate (on the basal 1/3-1/2). *Inflorescences*: Unisexual. *Male inflorescences*: Racemose; 3-15-flowered; sessile. *Male flowers*: Pedicel terete in cross section; 2-6(-16) mm long; antrorse-strigose; rarely subtended by a bracteole. Bracteole subtending pedicel very narrowly elliptic in outline; narrowly long cuneate at the base; acute or narrowly acute at the apex; 2-2.8 X ca 2.5 mm; sparsely antrorse-strigose at apex; glabrous at base. Hypanthium campanulate; 4.5-6.5 mm long; ca 3.5 mm in diam; sparsely short hirsute. Calyx lobes triangular to narrowly triangular in outline; narrowly acute at the apex; 1.6-2 X 0.8-1.1 mm; glabrate. Corolla infundibular; glabrous outside; glabrous inside. Corolla tube 0.5-1 mm long; 2-2.4 mm in diam; glabrous outside. Corolla lobes broadly ovate to shallowly ovate in outline; acute to broadly acute at the apex; 2.2-3 X 2.5-2.8 mm; glabrous outside;

puberulent inside. Stamens separating from the hypanthium ca 1.6 mm from the base of the hypanthium. Filaments radially compressed in cross section; ca 1.1 mm long; ca 0.5 mm wide; glabrous. Anther thecae 2.4-2.8 mm long; puberulous. Anther connective obovate; unilobate; obtuse at apex; puberulous at the apex; 0.6-1 X (0.4-)0.8-1.4 mm. Disc cylindrical; 0.6-0.9 mm long; 1.4-2.1 mm in diam. *Female inflorescences*: A solitary flower. *Female flowers*: Pedicel terete in outline; (1.6-)3-11 mm long; hirsute (to sparsely so); with nonbreakaway hairs; cylindrical. The lower 2/3 of hypanthium fused to the ovary. Fused portion of hypanthium ovoid; ca 3.8 mm long; ca 2 mm in diam; aculeate; glabrous. Aculei 0.8-1 mm long. Hyaline bristle at apex of aculeus 0.4-0.6 mm long; 0.8-1.5(-3) times as long as the opaque base. The upper 1/3 of hypanthium free from the ovary. Free portion of hypanthium ca 1.4 mm long; 1.4-2.4 mm in diam; not aculeate outside; sparsely hirsute outside; glabrous inside. Calyx lobes triangular in outline; narrowly acute at the apex; 1.7-2.2 X ca 1.2 mm; glabrate. Corolla puberulent outside; glabrous inside. Corolla tube present; 0.1-0.2 mm long; ca 1 mm in diam; glabrous outside. Corolla lobes ovate in outline; broadly acute at the apex; 1.2-1.4 X 0.7-1 mm; glabrous outside (with the midvein sparsely short strigose); glabrous inside. Staminodes absent. Style ca 0.4 mm long; ca 0.4 mm in diam; subtended by a circular disc. Stigma ca 1.8 mm long; ca 1.8 mm in diam; lobate; with 9 finger-like projections on the margin. Finger-like stigmatic projections 0.8-1 mm long; 0.3-0.5 mm in diam; smooth. *Fruit*: Not geocarpic, maturing above ground and readily visible. Pedicel sulcate in cross section; 0.7-2 cm long; hirsute (to sparsely so); with nonbreakaway hairs; cylindrical. Fruit bicolored with longitudinal stripes from base to apex; dark green; with yellow longitudinal stripes; ellipsoid (to broadly so), or ovoid (to broadly so); 1.8-3(-5) cm long; 1.5-2(-4) cm in diam; aculeate; glabrous; blunt at the apex. Aculei terete; with hyaline bristle persistent; 2.5-3.5 mm long; 0.7-1 mm in diam. Bristle 0.3-0.4 mm long. *Seeds*: Elliptic; 5.5-7 X 3-4 mm; ca 1.5 mm thick; unwinged. *Chromosome number*: 2n = 48 (Shimotsuma, 1965; Deakin, Bohn & Whitaker, 1971; Dane & Tsuchiya, 1976, 1979; Ramachandran, 1984; Yadava, Singh & Arya, 1984; Ramachandran & Narayan, 1985).

Distribution: Africa: South Africa, in the provinces of Cape, Transvaal, and Orange Free State.

Based upon its chloroplast DNA structure (Perl-Treves and Galun, 1985), *C. heptadactylus* is most closely related to *C. myriocarpus* and a member of ser. *Myriocarpi*. It is easily distinguished by its extremely narrow leaf lobes and dioecious condition.

5. **Cucumis kalahariensis** Meeuse, Bothalia 8(1): 70. 1962.—TYPES: SOUTH AFRICA. Cultivated at Roodeplaat from seed collected in Namibia at Nama Pan (26 Aug 1955), Apr 1957, *R. Story 5320* (lectotype: PRE!; isolectotypes: PRE!). BOTSWANA. 272 miles northwest of Molopolole, 27 Jun 1955, *R. Story 4986* (paratypes: BR! K! M! PRE! SRGH! [only tubers collected in the field, aerial parts grown out at Pretoria, South Africa]); 132 miles northwest of Molopolole, scattered throughout the Kalahari as far north as Gautscha Pan, 15 Jun 1955, *R. Story 4895* (paratypes, BR! K! M! [only fruit collected]); Kaotwe, 10 Apr 1930, *G. van Son s.n.* [Hb. Transvaal Museum, 28803] (paratype: PRE!).

Plants: Herbs; perennial; with a woody rootstock; tuberous; dioecious. Tubers 30-50 cm under ground; ellipsoid or obovoid; 8-12 cm long; 1.5-2 cm in diam; glabrous. Stems procumbent; sulcate; not aculeate; short pilose; with nonbreakaway hairs. Stem hairs 0.6-1.2 mm long (soon broken off). Nodes not geniculate. Internodes 2-5 or 12-22 cm long. *Leaves*: Petioles 0.5-4 cm long; sulcate; not aculeate; pubescence a single type on each petiole; short pilose (with the hair base enlarged); with nonbreakaway hairs. Leaf blades 3-palmately or

5-palmately lobed; with the margin serrate; broadly ovate in outline; subcordate to broadly acute at the base; without a basal sinus; acute to broadly acute at the apex; 4-6.5(-10) X 3-5.5 cm; 1.2-1.3(-2) times longer than wide; pilose on the upper surface; not aculeate on the veins below; pilose on the veins below; pilose on the intervenium below; with nonbreakaway hairs. Central leaf-blade lobe pinnatifid; narrowly elliptic to elliptic in outline; acute to broadly acute at the apex; 3-5.5(-8) X (0.5-)1-3 cm. Lateral leaf-blade lobes symmetrical; entire or pinnatifid; narrowly oblong in outline; acute at the apex; 0.7-2(-4.5) X 0.5-0.8(-1.5) cm. *Tendrils*: Present; solitary; simple; 1-2.5(-5) cm long; not aculeate; sparsely short pilose. *Inflorescences*: Unisexual. *Male inflorescences*: Fasciculate or a solitary flower; 1-3-flowered; sessile. *Male flowers*: Pedicel terete in cross section; 5-12 mm long; pilose; without bracteoles. Hypanthium infundibular; ca 8 mm long; ca 4 mm in diam; antrorse-strigose. Calyx lobes linear in outline; narrowly acute at the apex; ca 7.2 X ca 0.6 mm; sparsely antrorse-strigose. Corolla infundibular; puberulent outside; puberulent inside. Corolla tube ca 2.9 mm long; ca 8 mm in diam; puberulent outside. Corolla lobes elliptic in outline; acute at the apex; 9.6-11.2 X 6.4-8 mm; puberulent outside; puberulent inside. Stamens separating from the hypanthium ca 3.2 mm from the base of the hypanthium. Filaments radially compressed in cross section; ca 1.2 mm long; 0.4-0.8 mm wide; glabrous. Anther thecae ca 4 mm long; puberulent. Anther connective ovate; unilobate; obtuse at apex; papillate at the apex; 1.6-2 X ca 1 mm. Disc cylindrical; ca 1 mm long; ca 2.2 mm in diam. *Female inflorescences*: A solitary flower. *Female flowers*: Pedicel terete in outline; 4-10 mm long; pilose; with nonbreakaway hairs; cylindrical. The lower 2/3 of hypanthium fused to the ovary. Fused portion of hypanthium ellipsoid; 9.6-11 mm long; 5.6-6.4 mm in diam; aculeate; glabrous. Aculei soft; 1.6-2.4 mm long. Hyaline bristle at apex of aculeus 0.6-0.9 mm long; ca 0.6 times as long as the opaque base. The upper 1/3 of hypanthium free from the ovary. Free portion of hypanthium 4-6.4 mm long; ca 4.8 mm in diam; not aculeate outside; antrorse-strigose outside; puberulent inside. Calyx lobes linear in outline; narrowly acute at the apex; 5.9-6.4 X 0.6-0.8 mm; antrorse-strigose. Corolla puberulent outside; puberulent inside. Corolla tube present; ca 1.9 mm long; ca 5.6 mm in diam; puberulent outside. Corolla lobes elliptic in outline; acute at the apex; 9.6-11.2 X 4.6-5.4 mm; puberulent outside; puberulent inside. Staminodes present; separating from the free portion of the hypanthium ca 2.4 mm above the ovary; 0.8-1.6 mm long; 0.3-0.4 mm in diam; puberulent. Style ca 2.4 mm long; ca 1.1 mm in diam; subtended by a circular disc. Disc ca 0.8 mm long; ca 2.8 mm in diam. Stigma ca 3.8 mm long; ca 1.8 mm in diam; lobate; with 5 finger-like projections on the margin. Finger-like stigmatic projections ca 1.2 mm long; ca 0.8 mm in diam; papillate. *Fruit*: Not geocarpic, maturing above ground and readily visible. Pedicel terete in cross section; 1-3.5 cm long; pilose; with nonbreakaway hairs; cylindrical. Fruit bicolored with longitudinal stripes from base to apex; greenish white; with brownish purple longitudinal stripes; ellipsoid or cylindrical; 3-5.5 cm long; 2-3 cm in diam; aculeate; glabrous; blunt at the apex. Aculei laterally compressed near the base; with hyaline bristle lost; 1.6-2.8 mm long; 0.8-1.2 mm in diam. Bristle ca 0.4 mm long (when present). *Seeds*: Elliptic; 5.2-5.6 X 3-3.2 mm; ca 1.6 mm thick; unwinged.

   Distribution: Africa: central and northwestern Botswana and northeastern Namibia.

   Cucumis kalahariensis is found in the northern part of the Kalahari thornveld, in the southern areas of the zone transitional to Zambezian broad-leaved woodland as defined by White (1983). According to the herbarium labels, it has been encountered only on Kalahari sands. The tuberous roots are eaten by bushmen, and are an important element of their diet during the dry season.

   Story's first wild collections had only fruits and tubers. He grew out the tubers which produced only male plants. Sigmund Rehm, Division of Horticulture, grew plants at Roodeplaat, South Africa, from seeds in fruits collected by Story. Female and male plants, including fruit, were prepared as 4 specimens now at PRE and used as the *type* by Meeuse (1962). One is labeled as "type of female plant," another as "type of male plant," and the other 2 have no indication as type specimens. Since no single specimen is indicated as type, they are syntypes, and the specimen labeled "type of female plant" with flowering and fruiting stems is here chosen as lectotype.

Cucumis ser. **Angurioidei** Kirkbride, ser. nov.—TYPE: *Cucumis anguria* Linnaeus. Species 6-24.

A ser. *Myriocarpis* plantis maturis cucurbitacino *A*, provisis cum ser. *Myriocarpis* infertilibus, sect. *Melonis* fructuum crementum excitantibus, peroxidasi-4 allelomorphi *5* (Esquinas-Alcazar, 1977), sine chloroplasti acidi deoxyribonucleici PvuII-endonucleasi scissi loco scisso *20* (Perl-Treves and Galun, 1985) absimilis.

The circumscription of *Cucumis* ser. *Angurioidei* as a species group by Dane (1976, 1983), Esquinas-Alcazar (1977), Perl-Treves and Galun (1985), and Jeffrey (1980) was much broader and included the members of *Cucumis* ser. *Myriocarpi*. It is a group of cross-compatible species that are cross-incompatible with members of *Cucumis* ser. *Myriocarpi* and do stimulate fruit set in members of *Cucumis* ser. *Melo* (Raamsdonk, 1989). It is also supported by chemical and DNA characters summarized in the Key to the Subgenera, Sections, and Series of *Cucumis*. *Cucumis prophetarum* is additive and intermediate in its flavonoid patterns (Brown et al., 1969) between the two series having compound 19 in common with *Cucumis* ser. *Angurioidei* and compounds 11 and 15 in common with *Cucumis* ser. *Myriocarpi*. Many species have been placed in this series because of morphological similarity: *C. baladensis*, *C. carolinus*, *C. globosus*, *C. hastatus*, *C. insignis*, *C. jeffreyanus*, *C. meeusei*, *C. prolatior*, *C. rigidus*, *C. sacleuxii*, and *C. thulinianus*. Living material of these species is essential for the biosystematic research required to verify or clarify the positions of these species. *Cucumis* ser. *Angurioidei* is predominantly in the area covered by the *Flora of Tropical East Africa* (Jeffrey, 1967). Three species, *C. meeusei*, *C. rigidus*, and *C. zeyheri*, are found exclusively in southern Africa; *C. anguria* both in eastern and southern Africa; *C. prophetarum* in eastern Africa and extends through southwest Asia to India; and *C. pustulatus* occurs also in the southwest corner of the Arabian Peninsula, in Yemen, as well as in eastern Africa.

6. **Cucumis anguria** Linnaeus, *Sp. pl. ed. 1*, 1011. 1753.

Plants: Herbs; annual; without a woody rootstock; lacking tubers; monoecious. Stems procumbent or climbing; sulcate (to weakly so when older); not aculeate; hispid; with nonbreakaway hairs. Stem hairs 0.7-1.6(-3.3) mm long. Nodes not geniculate. Internodes (3-)5-13 cm long. *Leaves*: Petioles 6-13(-18) cm long; sulcate; not aculeate; pubescence a single type on each petiole; hispid to setose; with nonbreakaway hairs. Leaf blades deeply 3- or 5-palmately lobed, or trilobate (weakly); with the margin entire to serrate (rarely, regularly); ovate to broadly ovate to shallowly ovate in outline; cordate at the base; with a basal sinus; with the sinus 0.5-3 cm deep; acute to broadly acute at the apex; 3-12(-15) X 2.5-12(-15) cm; 0.9-1(-2) times longer than wide; hispidulous on the upper surface; not aculeate on the veins below; hispid on the veins below; hispidulous on the intervenium below; with nonbreakaway hairs. Central leaf-blade lobe entire; narrowly elliptic to broadly elliptic in outline; acute to broadly acute at the apex; 2-8(-10) X 1.5-7 cm. Lateral leaf-blade lobes with outer lobes asymmetrical and inner lobes symmetrical (when present); entire; elliptic to broadly elliptic in outline; obtuse or broadly acute at the apex; 1.5-7 X 1-5 cm. *Tendrils*: Present; solitary; simple; 3-6 cm long; not aculeate; setose (to sparsely so). *Inflorescences*: Unisexual. *Male inflorescences*: Reduced racemose; 3-10-flowered; pedunculate or sessile. Peduncle 0.1-2.5 cm long (when present). *Male flowers*: Pedicel terete in cross section; 4-30 mm long; hispidulous; without bracteoles. Hypanthium infundibular; 3.5-4 mm long; ca 2.8 mm in diam; hispidulous (to sparsely so). Calyx lobes narrowly triangular in outline; narrowly acute at the apex; 1.5-2.5 X 0.4-0.5 mm; sparsely hispidulous. Corolla infundibular; sparsely hispidulous outside; sparsely puberulent inside. Corolla tube ca 1.5 mm long; 3-3.5 mm in diam; sparsely hispidulous outside. Corolla lobes broadly ovate or broadly elliptic in outline; broadly acute at the apex; 4-6.5 X 3.5-5.6 mm; sparsely puberulent outside; sparsely puberulent inside. Stamens separating from the hypanthium 1.6-2.5 mm from the base of the hypanthium. Filaments radially compressed in cross section; 0.6-1 mm long; 0.2-0.4 mm wide; glabrous. Anther thecae 1.8-2 mm long; glabrous. Anther connective oblong to narrowly oblong; unilobate; obtuse at apex; papillate at the apex; 0.6-0.8 X 0.2-0.4 mm. Disc cylindrical; 0.6-0.8 mm long; 1-1.2 mm in diam. *Female inflorescences*: A solitary flower. *Female flowers*: Pedicel sulcate in outline (to weakly so); (1.5-)2-4.5(-7) mm long; hispid to setose; with nonbreakaway hairs; flaring upwards from a narrower base to a wider apex. The lower 2/3 of hypanthium fused to the ovary. Fused portion of hypanthium ellipsoid to cylindrical; 7-13 mm long; 4-6 mm in diam; densely to moderately aculeate; glabrous. Aculei soft; (0.7-)0.9-4 mm long. Hyaline bristle at apex of aculeus 0.5-1.1 mm long; 0.5-1 or 2-3 times as long as the opaque base. The upper 1/3 of hypanthium free from the ovary. Free portion of hypanthium 3-5 mm long; 3-5 mm in diam; not aculeate outside; hispidulous outside; glabrous inside. Calyx lobes narrowly triangular to linear in outline; narrowly acute at the apex; 1.8-4 X 0.4-0.6 mm; sparsely hispidulous. Corolla sparsely hispidulous outside; sparsely puberulent inside. Corolla tube present or absent; ca 1 mm long; ca 4 mm in diam; sparsely hispidulous outside. Corolla lobes obovate to broadly obovate in outline; acute to broadly acute at the apex (and apiculate); 6-8 X 4-5 mm; sparsely hispidulous outside; sparsely hispidulous inside. Staminodes present; separating from the free portion of the hypanthium 1.5-2.5 mm above the ovary; 0.4-1.2 mm long; ca 0.2 mm in diam; glabrous. Style 1.6-2.4 mm long; ca 0.6 mm in diam; subtended by a circular disc. Disc 0.6-1 mm long; 1.8-2.2 mm in diam. Stigma 2-2.5 mm long; 2-2.5 mm in diam; entire to sublobate. Finger-like stigmatic projections densely smooth. *Fruit*: Not geocarpic, maturing above ground and readily visible. Pedicel sulcate in cross section; (4-)6-14(-19) cm long; setose to hispid; with nonbreakaway hairs; flaring upwards from a narrower base to a wider apex. Fruit monocolored or bicolored with longitudinal stripes from base to apex (before ripe); yellowish green to yellow (light); with light green longitudinal stripes (when immature); ellipsoid to obovoid (very rarely); 2-7 cm long; 1.5-4 cm in diam; aculeate; glabrous; blunt at the apex. Aculei terete; with hyaline bristle lost; 0.8-1.9 or 4-10(-15) mm long; 0.8-1.4 mm in diam. *Seeds*: Elliptic; 5-6 X 2-2.5 mm; ca 1 mm thick; unwinged. *Chromosome number*: 2n = 24 (Kozuchov, 1930; Shimotsuma, 1965; Deakin, Bohn & Whitaker, 1971; Dane & Tsuchiya, 1976; Ramachandran, 1984; Singh & Yadava, 1984; Yadava, Singh & Arya, 1984; Nijs & Visser, 1985; Ramachandran & Narayan, 1985).

    Distribution: Africa: Angola, Botswana, Cape Verde Islands, Malawi, Mozambique, Namibia, South Africa, Sierra Leone, Swaziland, Tanzania, Zaire, Zambia, and Zimbabwe. Introduced: widely cultivated in the New World tropics and rarely in other areas of the world.

Key to the Varieties of *Cucumis anguria*

1. Apical hyaline bristle of aculei on hypanthium of female flower 2-3-times longer than the opaque base; aculei on fruit 0.8-1.9 mm long, appearing as small bumps or warts; leaf blade deeply 3- or 5-palmatilobate   *C. anguria* var. *anguria*

1. Apical hyaline bristle of aculei on hypanthium of female flower shorter than the opaque base; aculei on the fruit 4.0-10.0(-15.0) mm long, appearing as aculei; leaf blade deeply 3- or 5-palmatilobate or weakly trilobate, almost entire   *C. anguria* var. *longaculeatus*

6a. **Cucumis anguria** Linnaeus var. **anguria**—*Cucumis anguria* Linnaeus, Sp. pl. *ed. 1*, 1011. 1753.—TYPE (here designated): SWEDEN. Plant cultivated at Uppsala, Sweden (lectotype: LINN, sheet number 1152.6!).

Cucumis subhirsutus minor P. Browne, Civ. nat. hist. Jamaica ed. 2, 353. 1789.

Cucumis echinatus Moench, nom. superfl., Methodus 654. 1794.

Cucumis parviflorus Salisbury, nom. superfl., Prodr. stirp. Chap. Allerton 157. 1796.

Cucumis macrocarpus Wenderoth in Spix & C. Martius, Reise Bras. 2: 536. 1828.

Cucumis angurioides Roemer, Fam. nat. syn. monogr. 2: 79. 1846.—TYPE: Fl. Flum. Icon. 10: 85, 1831 (lectotype). Nom. nov. for *Cucumis anguria* Vell., Fl. Flum. Icon. 10: 85. 1831.

Cucumis erinaceus Naudin ex Huber, Catalogue pour le printemps 1864, p. 6. 1864.—TYPES: FRANCE. Muséum d'Histoire Naturelle, 1862, *C. Naudin s.n.* (lectotype: K!; isolectotypes: E! P).

Cucumis longipes Hooker filius in Oliver, Fl. trop. Afr. 2: 547. 1871.—TYPE: ANGOLA. Golungo Alto: Ad Quibine sob. Bumba, VI 1836, *F.M.J. Welwitsch 846* (lectotype: BM!; isolectotype: LISU!).—*Cucumis anguria* Linnaeus var. *longipes* (Hooker filius) Meeuse, Blumea, suppl., 4: 200. 1958.

Cucumis arada Linnaeus ex Naudin & Müller, Manuel de l'Acclimateur 230. 1887. I examined the *Cucumis* collections at P in 1987 and was unable to locate the type of *C. arada*. Presumably it was described from garden plants.

Cucumis anguria Linnaeus subsp. *cubensis* Gandoger, Bull. Soc. Bot. France 65: 28. 1918.—TYPE: CUBA. Pinar del Río: Punta Brava, 15 Nov 1904, *Baker & O'Donovan 3992* (holotype: LY!). *Cucumis cubensis* Gandoger, Bull. Soc. Bot. France 65: 28. 1918, indexing error.

Cucumis anguria Linnaeus subsp. *jamaicensis* Gandoger, Bull. Soc. Bot. France 65: 28. 1918.—TYPE: JAMAICA. Hope, 1 Jun 1898, *W. Harris 6967* (holotype: LY!). *Cucumis jamaicensis* Gandoger, Bull. Soc. Bot. France 65: 28. 1918, indexing error.

Leaves: Leaf blades deeply 3-palmately or 5-palmately lobed. *Female flowers*: Hyaline bristle at apex of aculeus 2-3 times as long as the opaque base. *Fruit*: Aculei 0.8-1.9 mm long (appearing as small bumps or warts on the fruit).

6b. **Cucumis anguria** Linnaeus var. **longaculeatus** Kirkbride, var. nov.—TYPE: NAMIBIA. Ovamboland: 20 km. south of Oshikango, 12 Mar 1973, *R. J. Rodin 9046* (holotype: K!; isotypes: M! MO! PRE!).

A *C. anguria* var. *anguria* floris feminei hypanthii aculeorum seta hyalina apicali breviore quam basi opaca, fructuum aculeis 4.0-10.0(-15.0) mm longis, folii lamina profunde palmatim 3- vel 5-lobata vel leviter trilobata absimilis.

Leaves: Leaf blades deeply 3- or 5-palmately lobed, or trilobate (weakly). *Female flowers*: Hyaline bristle at apex of aculeus 0.5-1 times as long as the opaque base. *Fruit*: Aculei 4-10(-15) mm long .

Cucumis anguria has been known since before 1650 as a cultivated plant of the New World (Piso and Marcgrave, 1648; Piso, 1658). It was assumed to be native to the West Indies, in particular to Jamaica (de Candolle, 1882), and consequently *Cucumis* was considered to be an African genus with one native species in the New World. Naudin (1859) was the first to suggest that *C. anguria* was introduced into the New World from Africa, but concluded that it was really native to the New World because of its early discovery in America and his inability to identify it with any known African species. Hooker (1870) believed that it had originated from a west African species of *Cucumis*, and that it had been so modified in cultivation that its progenitor was unknown. De Candolle (1882) agreed that it was introduced from Africa and presented two lines of evidence to support his conclusion: firstly, one of its common names among negroes in the French Antilles was "concombre marron." In this context, "marron" means "run wild" referring to an introduced plant or cultivar. Secondly, it was found only in areas of the Americas that were active in the slave trade; it did not occur inland or on the Pacific coast.

Meeuse (1958) cleared up the mystery by proposing that *C. anguria* was conspecific with *C. longipes* and demonstrating this through hybridization between the two taxa. He reduced the wild form to varietal status under *C. anguria* and stated that the cultivated variety was unknown from Africa. He accepted the hypothesis of Hooker (1870) and de Candolle (1882) that the cultigen had been introduced by slaves imported into the New World from west Africa. When slaves were collected in the interior of Africa, they were formed into groups and chained together for the march to the coast. Each was forced to carry a sack of course cloth with corn meal in it, their provisions for the trip (Spix and Martius, 1823-1831; Carreira, 1968; Fernandes, 1985). Undoubtedly anything edible encountered during the march was picked up to supplement the meager diet, and one of those things was probably *C. anguria* var. *longaculeatus*. Sometimes the sacks were taken aboard ship and carried to the New World with their contents, and thus *C. anguria* var. *longaculeatus* was carried to the New World.

The two varieties can be easily separated by the relative length of the apical hyaline bristle to the opaque basal portion of the aculei on the hypanthium of the female flower and fruits and by the length of the aculei on the fruits. The cultivated variety has the hyaline bristle longer than the opaque basal portion, and the aculei on the fruit remain relatively short. On the wild variety, the hyaline bristle is always shorter than the opaque basal portion, and the aculei on the fruit lengthen significantly during growth of the fruit and are always longer than those of the cultivated variety.

Jeffrey (1967) selected a lectotype for the cultivated variety, *C. anguria* var. *anguria*, in the Linnean herbarium. There are two sheets, LINN 1152.5 and 1152.6, identified as *C. anguria* and pinned together. Jeffrey did not indicate which sheet was the lectotype, so he actually designated lectosyntypes for the species. Number 1152.6 is sterile, and number 1152.5 has small, immature fruit, the largest being 1.2 cm long and 1 cm in diameter. The aculei on the fruit of number 1152.5 are approximately 2.0 mm long with the hyaline bristle 1/4 to 1/3 as long as the opaque base. Number 1152.6 is a member of variety *anguria*, and number 1152.5 is probably a member of variety *longaculeatus* with immature fruit. Also, number 1152.5 appears to be a post-1753 addition to the herbarium (C.E. Jarvis, pers. comm.), so I have chosen number 1152.6 as lectotype.

The wild variety was designated *C. anguria* var. *longipes* by Meeuse based upon the Hooker binomial *C. longipes* (1871). In the protologue, Hooker states, "Berry ... quite smooth and glabrous except for a very few scattered prickles." He cited Welwitsch's collections from Angola as representing the new species. Welwitsch made four collections of *C. anguria* in Angola; two are members of the cultivated variety (*824* [BM! LISU!]), *846* [BM! LISU!]), one is the wild variety (*847* [LISU!]), and the last is somewhat intermediate in length of aculei between the two varieties (*848* [BM! LISU!]). Only numbers *824* and *846* bear fruit corresponding to those described by Hooker. The fruit of *Welwitsch 846*, herein designated lectotype of *C. longipes*, clearly identify it as synonymous with the cultivated variety, and so its epithet can not be used for the wild variety. Therefore, I have named the wild variety *C. anguria* var. *longaculeatus*.

The leaf blade of *C. anguria* var. *longaculeatus* is variable, ranging from deeply 3- or 5-palmatilobate to weakly trilobate or almost entire. This variation is found throughout the range of the variety. In *C. anguria* var. *anguria*, the leaf blade is always deeply lobed and may have either three or five lobes.

Mabberley (1985) discovered many previously unrecognized epithets published in garden catalogues of Ch. Huber Frères et Cie., Antibes, France, in the last century, including *C. erinaceus*. Charles Naudin routinely sent to Huber seeds of the Cucurbitaceae grown in the gardens of the Muséum d'Histoire Naturelle, Paris. The description published by Huber is sufficient to indicate that the plants might be *C. anguria* var. *anguria*. I was unable to locate specimens annotated as *C. erinaceus* by Naudin at P, so to fix the identity of *C. erinaceus*, I selected a garden collection of *C. anguria* var. *anguria* made by Naudin in 1862 at the Muséum d'Histoire Naturelle, Paris, as lectotype.

7. **Cucumis sacleuxii** Paillieux & Bois, Bull. Soc. Natl. Acclim. France 37(8): 371. 1890.—TYPE: TANZANIA. Zanzibar: Massazine, alt. sea-level or 100 ft., 15 Jul 1961, *H.G. Faulkner 2865* (neotype: BR!; isoneotype: K!).

Oreosyce aspera Cogniaux, Pflanzenr. 66: 268. 1916.—TYPE: ZAIRE. Kivu: Forestier Central, Dundusana, Nov 1918, *M.G. Mortehan 720* (holotype: BR!).

Cucumis sativus Linnaeus var. *usambarensis* Zimmermann, Cucurbitac. 2: 179. 1922.—TYPE: TANZANIA. Lushoto District: Msituni, Aug 1920, *A. Zimmermann s.n.* (neotype (Jeffrey, 1967): EA).

Plants: Herbs; annual; monoecious. Stems procumbent or climbing; sulcate; not aculeate; pilose (hairs of 2 lengths, the longest evident without magnification and much longer than the others); with breakaway hairs. Stem hairs 0.3-4.4 mm long. Nodes not geniculate. Internodes 3-8(-16) cm long. *Leaves*: Petioles 2-8(-13) cm long; sulcate; not aculeate; pubescence a single type on each petiole; pilose (hairs of 2 lengths, the longest evident without magnification and much longer than the others); with breakaway hairs. Leaf blades pentalobate to trilobate; with the margin serrate; very broadly ovate in outline; cordate at the base; with a basal sinus; 1-2.5 cm deep; acute to broadly acute, or obtuse (rarely) at the apex; 3.5-9(-11) X 4-9(-11) cm; 0.8-1 times longer than wide; pilose on the upper surface; not aculeate on the veins below; pilose on the veins below (hairs of 2 lengths, the longest evident without magnification and much longer than the others); pilose on the intervenium below; with breakaway hairs. Central leaf-blade lobe entire; ovate to broadly ovate to shallowly ovate in outline; acute to broadly acute, or obtuse (rarely) at the apex; 2-4 X 2-6 cm (largest when leaf trilobate). Lateral leaf-blade lobes asymmetrical; entire; shallowly ovate in outline; acute to broadly acute, or obtuse (rarely) at the apex; 1-1.5 X 2-3 cm. *Tendrils*: Present; solitary; simple; 3-8 cm long; not aculeate; short pilose. *Inflorescences*: Unisexual. *Male inflorescences*: A solitary flower or fasciculate; 1-4-flowered; sessile. *Male flowers*: Pedicel terete in cross section; 4-15 mm long; pilose; without bracteoles. Hypanthium infundibular, or campanulate; ca 4.8 mm long; 3.2-3.6 mm in diam; pilose. Calyx lobes linear or narrowly triangular in outline; narrowly acute at the apex; 3-3.8 X 0.2-0.5 mm; pilose. Corolla infundibular; densely puberulent and (sparsely) pilose outside; puberulent inside. Corolla tube 0.5-1.2 mm long; ca 4.8 mm in diam; densely puberulent and pilose (sparsely) outside. Corolla lobes obovate in outline; broadly acute, or obtuse and mucronate at the apex; 6.4-7.6 X 4.8-6 mm; densely puberulent and pilose (sparsely) outside; puberulent inside. Stamens separating from the hypanthium ca 2.4 mm from the base of the hypanthium. Filaments radially compressed in cross section; ca 0.6 mm long; 0.3-0.4 mm wide; glabrous. Anther thecae 2.4-2.8 mm long; puberulent. Anther connective obovate; unilobate; obtuse at apex; papillate at the apex; 1.6-2 X ca 1 mm. Disc cylindrical; 1.2-2 mm long; 2.4-3.3 mm in diam. *Female inflorescences*: A solitary flower. *Female flowers*: Pedicel sulcate in outline; 35-60 mm long; pilose; with breakaway hairs; cylindrical. The lower 3/4 of hypanthium fused to the ovary. Fused portion of hypanthium ellipsoid to cylindrical; ca 9.6 mm long; ca 4 mm in diam; aculeate; glabrous. Aculei 0.7-1.4 mm long. Hyaline bristle at apex of aculeus 0.5-1 mm long; 1.5-2.5 times as long as the opaque base. The upper 1/4 of hypanthium free from the ovary. Free portion of hypanthium ca 4.4 mm long; ca 3.2 mm in diam; not aculeate outside; pilose outside. Calyx lobes linear in outline; narrowly acute at the apex; ca 2.4 X ca 0.2 mm; pilose. Corolla puberulent and pilose (sparsely) outside. Corolla tube present; ca 1 mm long; ca 4 mm in diam; puberulent and pilose (sparsely) outside. Corolla lobes obovate in outline; broadly acute and mucronate at the apex; 6.8-7.2 X ca 4 mm; puberulent and (sparsely) pilose outside; puberulent inside. *Fruit*: Not geocarpic, maturing above ground and readily visible. Pedicel terete in cross section; 3.5-7 cm long; pilose; with breakaway hairs; cylindrical. Fruit bicolored with longitudinal stripes from base to apex; dark green; with light green or with white longitudinal stripes; ellipsoid to cylindrical; 4.5-8 cm long; 2-5 cm in diam; sparsely aculeate; glabrous; blunt at the apex. Aculei terete; with hyaline bristle lost; 0.5-1 mm long; 0.5-1.1 mm in diam. *Seeds*: Elliptic; 5.6-6.6 X 2.5-2.9 mm; 1.1-1.4 mm thick; unwinged. *Chromosome number*: 2n = 24 (Kozuchov, 1930).

Distribution: Africa: Kenya, Madagascar, Tanzania (including Zanzibar), Uganda, and Zaire.

Cucumis sacleuxii is restricted to eastern tropical Africa, including western Zaire, and Madagascar. It was first collected at Mandera, Tanzania, by R.P. Sacleux, who sent seeds to the Muséum d'Histoire Naturelle, Paris. Pailleux and Bois obtained some of the seeds from the Muséum and grew them out. They concluded that it was a new species, and Charles Naudin concurred (Pailleux & Bois, 1890). Jeffrey (1967) postulated that their holotype might be deposited at P, but I was unable to locate the holotype of *C. sacleuxii* there. Probably it was grown out, studied, and described as a living plant in the gardens of the Muséum, and an herbarium specimen was not prepared. Therefore, I have here designated as neotype a specimen with two female flowers and a fruit, *Faulkner 2865* (BR).

The hairs of *C. sacleuxii* are distinctive. They are the *non-glandular, multicellular, conical, simple* type of Inamdar and Gangadhara (1975) and Inamdar et al. (1990). They consist of a multicellular foot and a straight, uniseriate apex. The foot cells are smaller and have thinner walls than those of the apex. The stem, petioles, and leaf blades are pilose with two layers of hairs. A denser, lower layer covers all the organs, and there are relatively longer, easily distinguished hairs scattered over the organs forming the second or upper layer of hairs. The cells of the foot are fragile and rupture easily separating the apex from the plant. Jeffrey (1967) referred to these as *irritant* hairs; he presumably theorized that they are a defense mechanism. If this is correct, the mechanism is mechanical rather than chemical.

The lower, fused portion of the hypanthium on female flowers is scabrous. The basic structure of the hairs is the same as those on the stems, petioles, and leaves, i.e., a multicellular foot and a straight, unseriate apex. The walls of the foot cells are thick and resistant, and the cells of the apex are much smaller. The apex is lost during fruit maturation, and the foot is persistent on the mature fruit.

8. **Cucumis carolinus** Kirkbride, sp. nov.—TYPES: KENYA. District of Wajir, 4 km SW of Habaswein, 240 m, 27 Apr 1978, *M.G. Gilbert & M. Thulin 1116* (holotype: K!; isotype: UPS!); Northern Frontier Province, 30 km S of Marsabit on Isiolo road near southern edge of Marsabit Massif, 900 m, 24 Nov 1977, *S. Carter & B. Stannard 697* (paratype: K!); District of Isiolo, Isiolo-Wajir, Ewasso Nyiro flood plain, 35 km E of Modo Gash, sides of Murram causeway, 240 m, 9 Dec 1977, *B. Stannard & M.G. Gilbert 895* (paratype: K!). ETHIOPIA. Bircot, 2 Nov 1956, *C. Ashall CA4* (paratype: K!). Fig. 1.

A *Cucumeris* speciebus ceteris cujusque floris feminei hypanthio libero ad orem 7 mm diametro, corollae tubo ad orein 7-8 mm diametro, staminodiis a corollae tubo 3 mm supra ovarium secedentibus, stylo 3 mm alto et 1.4 mm diametro, disco 1.2 mm alto, seminibusque 1.8 mm latis differt.

Plants: Herbs; monoecious. Stems procumbent; sulcate; not aculeate; short retrorse-strigose; with nonbreakaway hairs. Stem hairs 0.2-0.6 mm long. Nodes not geniculate. Internodes 5.5-7 cm long. *Leaves*: Petioles 5-9 cm long; sulcate; not aculeate; pubescence a single type on each petiole; short retrorse-strigose or antrorse-strigose (rarely); with nonbreakaway hairs. Leaf blades 3-palmately lobed; with the margin serrate; broadly ovate in outline; cordate at the base; with a basal sinus; 0.4-1 cm deep; broadly acute at the apex; 6-7 X 5-5.5 cm; 1.1-1.3 times longer than wide; antrorse-strigose or scabrous on the upper surface; not aculeate on the veins below; antrorse-strigose on the veins below; scabrous or antrorse-strigose on the intervenium below; with nonbreakaway hairs. Central leaf-blade lobe entire; very broadly ovate to broadly ovate, or broadly elliptic in outline; broadly acute at the apex; 3-4 X 3.2-3.5 cm. Lateral leaf-blade lobes asymmetrical; entire; very shallowly ovate to shallowly triangular in outline; broadly acute at the apex; 1-2 X 3-4 cm. *Tendrils*: Present; solitary; simple; 3-7 cm long; not aculeate; glabrate or retrorse-strigose (sparsely). *Inflorescences*: Unisexual. *Male inflorescences*: Racemose or a solitary flower; 1- or 2-flowered; sessile. *Male flowers*: Pedicel sulcate in cross section; 65-115(-130) mm long; glabrate; without bracteoles. Hypanthium infundibular, or campanulate; 7.2-8.8 mm long; 4-5.6 mm in diam; glabrate or hispidulous. Calyx lobes narrowly triangular in outline; narrowly acute at the apex; 2-3.2 X 0.5-0.8 mm; glabrous to glabrate, or hispidulous (sparsely). Corolla infundibular; puberulent outside (to sparsely so); puberulent inside (to sparsely so). Corolla tube 2.4-3.2 mm long; ca 8 mm in diam; puberulent outside (to sparsely so). Corolla lobes broadly elliptic in outline; obtuse and mucronate at the apex; 16-20 X 13-17 mm; puberulent (to sparsely so) outside; puberulent (to sparsely so) inside. Stamens separating from the hypanthium 1-2 mm from the base of the hypanthium. Filaments radially compressed in cross section; ca 1.2 mm long; 0.5-0.8 mm wide; glabrous. Anther thecae 3.2-4 mm long; puberulent. Anther connective obovate; unilobate; obtuse at apex; papillate at the apex; ca 2 X 0.8-1 mm. Disc cylindrical; ca 0.8 mm long; ca 2 mm in diam. *Female inflorescences*: A solitary flower. *Female flowers*: Pedicel sulcate in outline; 10-100 mm long; glabrate or hispidulous; with nonbreakaway hairs; cylindrical. The lower 2/3 of hypanthium fused to the ovary. Fused portion of hypanthium ellipsoid; ca 15 mm long; ca 8 mm in diam; sparsely aculeate; glabrous. Aculei soft; 1.2-2.2 mm long. Hyaline bristle at apex of aculeus ca 0.4 mm long; 0.2-0.3 times as long as the opaque base. The upper 1/3 of hypanthium free from the ovary. Free portion of hypanthium ca 8 mm long; ca 7 mm in diam; not aculeate outside; sparsely antrorse-strigose outside; puberulent inside. Calyx lobes narrowly triangular in outline; narrowly acute at the apex; 2.4-3.2 X 0.8-1.2 mm; glabrate. Corolla puberulent outside; puberulent inside. Corolla tube present; 2-3 mm long; 7-8 mm in diam; puberulent outside. Corolla lobes broadly elliptic in outline; obtuse and mucronate at the apex; 15-20 X 13-16 mm; puberulent outside; puberulent inside. Staminodes

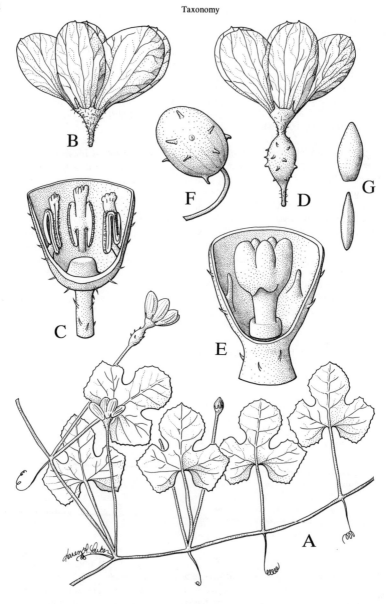

Fig. 1. *Cucumis carolinus*. A. Flowering stem with leaves, x ½. B. Male flower, x 1. C. Male flower, opened to show stamens and disc, x 5. D. Female flower, x 1. E. Female flower, opened to show disc, style, stigma, and staminodes, x 2½. F. Fruit, x ½. G. Seed, face and side views, x 5. (*Carter & Stannard 697* (K), B, C; *Gilbert & Thulin 1116* (K), A, D, E, F, G.)

present; separating from the free portion of the hypanthium ca 3 mm above the ovary; 1-1.2 mm long; 0.4-0.6 mm in diam; glabrous. Style ca 3 mm long; ca 1.4 mm in diam; subtended by a circular disc. Disc ca 1.2 mm long; ca 2.2 mm in diam. Stigma ca 3.6 mm long; ca 4 mm in diam; lobate; with 6 finger-like projections on the margin. Finger-like stigmatic projections ca 2 mm long; ca 1.2 mm in diam; papillate. *Fruit*: Not geocarpic, maturing above ground and readily visible. Pedicel sulcate in cross section; 2.5-6 cm long; glabrate; with nonbreakaway hairs; cylindrical. Fruit 5-5.5 cm long; 3-3.5 cm in diam; sparsely aculeate; glabrous; blunt at the apex. Aculei terete; with hyaline bristle lost; 3.6-6.4 mm long; 1.6-2 mm in diam. *Seeds*: Elliptic; 4-4.2 X ca 1.8 mm; 0.9-1 mm thick; unwinged.

    Distribution: Africa: Ethiopia and Kenya.

    Cucumis carolinus is known from only four collections, three from northeastern Kenya and one from eastern Ethiopia, between 240 and 900 m elevation. The description and figure were prepared principally from the Kenyan collections. The single Ethiopian collection, *Ashall CA4*, has denser pubescence and denser aculei on the fused portion of the female-flower hypanthium. Probably the fruit will also have denser aculei. The full range of variation in this species will not be understood until additional collections are available from southeastern Ethiopia.

    Charles Jeffrey (pers. comm.) recognized this as a new species, but did not name it because the fruits and seeds were unknown to him. Its epithet honors him and his many contributions to the systematics of Cucurbitaceae. The epithet *jeffreyanus* was not used because M. Thulin (1991) has already published a *C. jeffreyanus* from Somalia.

9. **Cucumis dipsaceus** Ehrenberg ex Spach, Hist. nat. vég. 6: 211. 1838.—TYPE: SAUDI ARABIA. Wadi Kamme east of al-Qunfidha, late February, 1825, *G. Ehrenberg & Hemprich s.n.* (holotype: B [destroyed (Jeffrey, 1967)]; lectotype: MPU!).

    Plants: Herbs; annual; without a woody rootstock; lacking tubers; monoecious. Stems procumbent or climbing; sulcate; not aculeate; on the ridges hispid and in the grooves hispidulous (with the longer hairs 0.8-1.5 mm long and the shorter ones ca 0.3 mm long); with nonbreakaway hairs. Stem hairs ca 0.3 or 0.8-1.5 mm long. Nodes not geniculate. Internodes 2-6(-9) cm long. *Leaves*: Petioles 1.5-5(-14) cm long; sulcate; not aculeate; pubescence a single type on each petiole; weakly hispidulous to hispid; with nonbreakaway hairs. Leaf blades entire or trilobate (rarely broadly); with the margin regularly serrate to entire; ovate to broadly ovate in outline; cordate at the base (sometimes with the lobes overlapping); with a basal sinus; 0.5-2(-3) cm deep; broadly acute or obtuse (rarely) at the apex; 3-7.5(-12.5) X 2-7(-12) cm; 1.1-1.5 times longer than wide; hispidulous, or hispidulous and hispid on the upper surface; not aculeate on the veins below; hispid on the veins below; hispidulous on the intervenium below; with nonbreakaway hairs. Central leaf-blade lobe entire or pinnatifid (rarely); broadly ovate to very broadly ovate to shallowly ovate, or narrowly elliptic (rarely) in outline; broadly acute or obtuse (rarely) at the apex; 2-4(-6.5) X 1.5-3.5(-7) cm. Lateral leaf-blade lobes asymmetrical; entire; shallowly ovate to very shallowly ovate, or narrowly elliptic (rarely) in outline; obtuse at the apex; 0.5-1.5(-3) X (0.5-)2-3(-5) cm. *Tendrils*: Present; solitary; simple; 1.5-6 cm long; not aculeate; basally hispidulous and apically glabrate. *Inflorescences*: Unisexual. *Male inflorescences*: A solitary flower or fasciculate; 1-5-flowered; sessile. *Male flowers*: Pedicel terete in cross section; 5-20 mm long; hispidulous; without bracteoles. Hypanthium narrowly infundibular; 3.6-5.2 mm long; 2.4-2.8 mm in diam; hispidulous. Calyx lobes narrowly oblong to linear in outline; narrowly acute at the apex; 1.6-4 X 0.1-0.3 mm; sparsely hispidulous. Corolla infundibular; sparsely hispidulous outside; glabrous inside. Corolla tube 1-1.5 mm long; ca 3 mm in diam; sparsely hispidulous outside. Corolla lobes obovate in outline; acute (and apiculate) at the apex; 5.5-8.5 X 3-5 mm; sparsely hispidulous outside; glabrous inside. Stamens separating from the hypanthium 2-2.8 mm from the base of the hypanthium. Filaments terete in cross section; 0.4-0.8 mm long; ca 0.2 mm wide (in diam); glabrous. Anther thecae ca 2.5 mm long; glabrous. Anther connective transversely broadly oblong; unilobate; obtuse at apex; weakly fimbriate at the apex; 0.6-0.8 X 0.6-1 mm. Disc cylindrical; ca 1.2 mm long; ca 1.5 mm in diam. *Female inflorescences*: A solitary flower. *Female flowers*: Pedicel sulcate (to weakly so) in outline; 5-15 mm long; hispid and hispidulous, or hispidulous; with nonbreakaway hairs; cylindrical. The lower 3/4 of hypanthium fused to the ovary. Fused portion of hypanthium ellipsoid; 8-19 mm long; 4-8 mm in diam; densely aculeate; glabrous. Aculei soft; 1.5-2.5 mm long. Hyaline bristle at apex of aculeus 1-1.4 mm long; 2-3 times as long as the opaque base. The upper 1/4 of hypanthium free from the ovary. Free portion of hypanthium 4-10 mm long; 3-5 mm in diam; not aculeate outside; hispidulous outside; glabrous inside. Calyx lobes linear in outline; narrowly acute at the apex; 4.8-5.6(-11) X 0.4-0.6(-1) mm; sparsely hispidulous. Corolla sparsely hispidulous outside; glabrous inside. Corolla tube present; ca 1.5 mm long; ca 2.5 mm in diam; sparsely hispidulous outside. Corolla lobes obovate in outline; acute at the apex; 6.5-15 X 3-8.5 mm; sparsely hispidulous outside; glabrous inside. Staminodes present; separating from the free portion of the hypanthium ca 1.2 mm above the ovary; 0.4-0.8 mm long; ca 0.3 mm in diam; glabrous. Style ca 1.5 mm long; ca 0.6 mm in diam; subtended by a circular disc. Disc ca 0.8 mm long; ca 1.8 mm in diam. Stigma ca 2.5 mm long; ca 2.8 mm in diam; lobate; with 5 finger-like projections on the margin. Finger-like stigmatic projections 0.8-1 mm long; ca 0.4 mm in diam; smooth. *Fruit*: Not geocarpic, maturing above ground and readily visible. Pedicel sulcate in cross section; 1-3(-4) cm long; hispid, or hispid and hispidulous; with nonbreakaway hairs; cylindrical. Fruit monocolored; pale yellow; ellipsoid to globose; 3-6.5 cm long; 2.5-4 cm in diam; densely aculeate; glabrous; blunt at the apex. Aculei terete; with hyaline bristle persistent; 4-6.4 mm long; ca 0.2 mm in diam. Bristle 1-1.5 mm long. *Seeds*: Elliptic; 4-5 X ca 2 mm; ca 1 mm thick; unwinged. *Chromosome number*: 2n = 24 (Kozuchov, 1930; Deakin, Bohn & Whitaker, 1971; Ramachandran, 1984; Singh & Yadava, 1984; Yadava, Singh & Arya, 1984; Nijs & Visser, 1985; Ramachandran & Narayan, 1985).

 Distribution: Africa: Ethiopia, Kenya, Somalia, Tanzania, Uganda, and possibly native to Sudan and southern Egypt. Introduced: widely adventive in the tropics.

 Evidence from chromosome pairing and pollen fertility of hybrids shows that *C. dipsaceus* is closest to *C. prophetarum* and *C. zeyheri* (Singh and Yadava, 1984). It is easily distinguished from other members of the genus by its densely aculeate fruit.

10. **Cucumis prophetarum** Linnaeus, Cent. pl. I 33. 1755.

 Plants: Herbs, or subshrubs; perennial; with a woody rootstock; lacking tubers; monoecious. Stems procumbent or climbing; sulcate; not aculeate; retrorse-strigose (to sparsely so), hispidulous (to sparsely so), or antrorse-strigose (to sparsely so); with nonbreakaway hairs. Stem hairs 0.3-0.8 mm long. Nodes not geniculate. Internodes (0.3-)1-7(-9) cm long. *Leaves*: Petioles 0.6-6.5 cm long; sulcate; not aculeate; pubescence a single type on each petiole; retrorse-strigose (to sparsely so), hispidulous (to sparsely so), or antrorse-strigose (to sparsely so); with nonbreakaway hairs. Leaf blades 3- or 5-palmately lobed, trilobate (weakly, sometimes, and rarely entire [*Lunt 112*, Libeh, Saudi Arabia]), or entire (almost); with the margin entire to serrate; ovate to broadly ovate to very broadly ovate in outline; cordate (to weakly so), truncate, or obtuse at the base; with or without a basal sinus; 0.5-1(-2) cm deep; acute to broadly acute, or obtuse at the apex; 2-7(-14) X 1.5-6.5(-8) cm; 0.8-1.6 times longer than wide; hispidulous (sparsely to densely so) on the upper surface; not aculeate on the veins below; hispidulous (sparsely to densely so) on the veins below; hispidulous (sparsely to densely so) on the intervenium below; with nonbreakaway hairs. Central leaf-blade lobe entire or pinnatifid; elliptic to broadly elliptic, or narrowly elliptic in outline; acute to broadly acute, or obtuse at the apex; 1.2-4.5(-12) X 0.4-2(-3) cm. Lateral leaf-blade lobes symmetrical or asymmetrical; entire or pinnatifid; narrowly elliptic or transversely elliptic in outline; acute to broadly acute, or obtuse at the apex; 0.3-2(-4.5) X 0.3-1.5(-2) cm. *Tendrils*: Present; solitary; simple; 1-3(-8) cm long; not aculeate; basally, sparsely retrorse-strigose, antrorse-strigose, or hispidulous. *Inflorescences*: Unisexual. *Male inflorescences*: A solitary flower, fasciculate, or racemose (rarely); 1-7-flowered; sessile. *Male flowers*: Pedicel terete in cross section; 3-8(-12) mm long; hispidulous (to sparsely so), retrorse-strigose, or antrorse-strigose; without bracteoles. Hypanthium infundibular; 2.8-4.4 mm long; 2-2.9 mm in diam; hispidulous. Calyx lobes oblong to narrowly oblong, or broadly triangular in outline; acute to narrowly acute at the apex; 0.4-1.6 X 0.2-0.5 mm; hispidulous (to sparsely so). Corolla infundibular; hispidulous (to sparsely so) outside; glabrous inside. Corolla tube 1-1.6(-2) mm long; 3-4 mm in diam; sparsely hispidulous to hirsute outside. Corolla lobes elliptic, or ovate to broadly ovate in outline; obtuse at the apex; 2.4-5.6 X 1-3.2 mm; sparsely hispidulous outside; glabrous inside. Stamens separating from the hypanthium ca 2 mm from the base of the hypanthium. Filaments terete in cross section; ca 0.8 mm long; 0.2-0.3 mm wide; glabrous. Anther thecae 2-2.8 mm long; minutely hispidulous or glabrous. Anther connective narrowly oblong to oblong; unilobate; obtuse at apex; papillate at the apex; 1-1.2 X 0.2-0.6 mm. Disc cylindrical; 1-1.2 mm long; 1.4-1.8 mm in diam. *Female inflorescences*: A solitary flower. *Female flowers*: Pedicel terete in outline; 5-40 mm long; hispidulous (to sparsely so), antrorse-strigose, or retrorse-strigose; with nonbreakaway hairs; cylindrical. The lower 2/3 of hypanthium fused to the ovary. Fused portion of hypanthium ellipsoid (to narrowly so) or ovoid; 4-8 mm long; 2.4-4 mm in diam; aculeate (to densely so); glabrous or puberulent. Aculei soft or stiff; 0.6-2.5 mm long. Hyaline bristle at apex of aculeus 0.4-1.2 mm long; 1.5-2 times as long as the opaque base. The upper 1/3 of hypanthium free from the ovary. Free portion of hypanthium 2-4.8 mm long; 2-4 mm in diam; not aculeate outside; hispidulous outside; glabrous inside. Calyx lobes narrowly oblong to linear, or narrowly obovate in outline; acute to narrowly acute at the apex; 1.2-4.2 X 0.3-0.6 mm; hispidulous. Corolla hispidulous outside; glabrous inside. Corolla tube present; 0.4-2 mm long; 3-4.4 mm in diam; hispidulous outside. Corolla lobes elliptic or ovate in outline; obtuse and mucronate, or acute (and apiculate, rarely) at the apex; 2.4-7.6 X 2-4 mm; hispidulous outside; sparsely hispidulous inside. Staminodes present; separating from the free portion of the hypanthium 1.6-2 mm above the ovary; 0.5-0.7 mm long; ca 0.2 mm in diam; glabrous. Style ca 1.2 mm long; 0.5-0.7 mm in diam; subtended by a circular disc. Disc 0.3-0.8 mm long; 1.2-1.8 mm in diam. Stigma 1.8-2 mm long; 1.6-2.4 mm in diam; lobate; with 5 finger-like projections on the margin. Finger-like stigmatic projections 1.2-1.6 mm long; 0.4-1 mm in diam; papillate. *Fruit*: Not geocarpic, maturing above ground and readily visible. Pedicel terete in cross section; 1-3.5(-6.5) cm long; sparsely retrorse-strigose, antrorse-strigose, or hispidulous; with nonbreakaway hairs; cylindrical. Fruit bicolored with longitudinal stripes from base to apex; green; with light green or white longitudinal stripes; ellipsoid (to broadly so) or globose; 2.5-4.5(-5) cm long; 1.2-4 cm in diam; aculeate; glabrous or hispidulous (sometimes); blunt at the apex. Aculei terete; with hyaline bristle persistent or lost; 1-8(-12) mm long; 0.6-1.4(-2) mm in diam. Bristle 0.4-0.6 mm long (when present on subsp. *dissectus*). *Seeds*: Elliptic; 4.3-6 X 2-2.6 mm; 1-1.3 mm thick; unwinged. *Chromosome number*: 2n = 24 (Kozuchov, 1930; Shimotsuma, 1965; Deakin, Bohn & Whitaker, 1971; Ramachandran, 1984; Singh & Yadava, 1984; Yadava, Singh & Arya, 1984; Nijs & Visser, 1985; Ramachandran & Narayan, 1985).

    Distribution given under the subspecies.

Key to the Subspecies of *Cucumis prophetarum*

1. Plants gray green; fruit pedicel 1-2.5(-3) cm long; fruit 2-4 cm in diam;
   seeds 5-6 X 2.4-2.6 mm   *C. prophetarum* subsp. *prophetarum*
1. Plants bright green; fruit pedicel 2-3.5(-6.5) cm long; fruit 1.2-2.5 cm
   in diam; seeds 4.3-5 X 2-2.2 mm   *C. prophetarum* subsp. *dissectus*

10a. **Cucumis prophetarum** Linnaeus subsp. **prophetarum**—TYPE: MIDDLE EAST. *F. Hasselquist s.n.* (holotype: LINN!; isotype: UPS).—*Cucumis foetidus* Salisbury, nom. superfl., Prodr. stirp. Chap. Allerton 156. 1796.
   Cucumis rheedei Kosteletzky, Allg. med.-pharm. Fl. 2: 738. 1833.—TYPE: Rheede, Hort. malab. 8: 21-22, *t. 11*. 1688. (Nicolson et al., 1988).
   Cucumis grossularoides Hort. in Steudel, Nomencl. bot, ed. 2, 1: 451. 1840. Invalid, Steudel cited it as a synonym of *C. prophetarum* Linnaeus, and later authors have cited it as a synonym of *C. myriocarpus*.
   Cucumis arabicus Delile, nom. nud., Index seminum horti regii botanici Monspeliensis 2. 1843.
   Cucumis amarus Stocks ex Naudin, pro syn., Ann. Sci. Nat., ser. 4, 11: 14. 1859.
   Cucumis anguinus Anderson, pro syn., Jour. Proc. Linnean Soc., suppl., 5: 19. 1860.
   Cucumis mascatensis Gandoger, Bull. Soc. Bot. France 65: 28. 1918.—TYPES: OMAN. Masqat [Maskat or Mascate], 24 Jan 1893, *A. Bornmueller 358* (lectotype: LY!; isolectotypes: BM!, K!, LE!, S!, W!), *P.M.R. Aucher-Eloy 4503* (syntypes: BM!, K!).

Plants: Internodes (0.3-)1-6 cm long. *Leaves*: Petioles 0.6-4 cm long. Leaf blades cordate to subcordate, or truncate at the base; with or without a basal sinus; 2-5 X 1.5-4 cm. *Male flowers*: Pedicel 3-5 mm long. Calyx lobes narrowly oblong in outline. Anther connective 0.5-0.6 mm. *Female flowers*: Pedicel 5-20 mm long. Aculei 0.6-1.6 mm long. Hyaline bristle at apex of aculeus 0.4-0.9 mm long; 1.3-2 times as long as the opaque base. Corolla lobes ca 7.6 X ca 4 mm. *Fruit*: Pedicel 1-2.5(-3) cm long. Fruit 2.5-4(-5) cm long; 2-4 cm in diam; aculeate; glabrous or hispidulous. Aculei 1-7 mm long; 0.6-1 mm in diam. *Seeds*: 5-6 X 2.4-2.6 mm.
   Distribution: Africa: Egypt, Mali, Mauritania, northern Nigeria, Senegal, Somalia, and Sudan. Southwest Asia: Iran, Iraq, Israel, Oman, Qatar, Saudi Arabia, Socotra, South Yemen, Syria, United Arab Emirates, and Jordan. Asia: northeastern India and Pakistan.

10b. **Cucumis prophetarum** Linnaeus subsp. **dissectus** (Naud.) C. Jeffrey, Kew Bull. 15(3): 351. 1962.—TYPE: ETHIOPIA. Golleb, Tacase, 1,000 m., 28 Feb 1853, *G.H.W. Schimper 1201* (lectotype: P!; isolectotypes: LE!, MPU!, P!, S!).—*Cucumis figarei* Delile ex Naudin var. *dissectus* Naudin, Ann. Sci. Nat., ser. 4, 11: 16. 1859. *Cucumis ficifolius* A. Richard var. *dissectus* (Naudin) Cogniaux, Monogr. phan. 3: 494. 1881.
   Cucumis figarei Delile ex Naudin var. *echinophorus* Naudin, Ann. Sci. Nat., sér. 4, 11: 16. 1859.—TYPE: YEMEN. Mont. Maamora, 21 Oct 1837, *P.E. Botta s.n.* (lectotype, P!; isolectotype, P!).—*Cucumis chrysocomus* Schumacher var. *echinophorus* (Naudin) Hiern, *Cat. afr. pl.* 1(2): 396. 1899.
   Cucumis pustulatus Hooker filius var. *echinophorus* A. Terr., Ann. Ist. Bot. Roma 5: 108. 1894.—TYPES: ETHIOPIA. Ertra, Amfile Bay, 22-23 Mar 1892, *A. Terracciano 77* (syntype: RO); *186* (syntype: RO [photograph K]).

Cucumis lyratus Zimmermann, Cucurbitac. 2: 178. 1922.—TYPE: TANZANIA. Lushoto District: Buiko, Dec 1919, *A. Zimmermann s.n.* (holotype).
Cucumis nigristriatus Zimmermann, Cucurbitac. 2: 178. 1922.—TYPE: TANZANIA. Pare District: Same, Aug 1920, *A. Zimmermann s.n.* (holotype).
Cucumis halabarda Chiovenda, Flora Somalia 1: 185. 1929.—TYPES. SOMALIA. Obbia, pozzo di Garbauen, 25 Apr 1927, *N. Puccioni & G. Stefanini 407* (syntype: FT [photograph K, no. 4839]); tra Scermarca Hassan e Tobungab, 4 May 1927, *N. Puccioni & G. Stefanini 582* (syntype: FT [photograph K, no. 4879]).

Plants: Internodes 4-7(-9) cm long. *Leaves*: Petioles 1.5-6.5 cm long. Leaf blades cordate at the base; with a basal sinus; 2-7(-14) X 1.8-6.5(-8) cm. *Male flowers*: Pedicel 4-8(-12) mm long. Calyx lobes broadly triangular, or oblong to narrowly oblong in outline. Anther connective 0.2-0.4 mm. *Female flowers*: Pedicel 20-40 mm long. Aculei 1.3-2.5 mm long. Hyaline bristle at apex of aculeus 1-1.2 mm long; 0.6-1.1 times as long as the opaque base. Corolla lobes 2.4-4.4 X 2-2.8 mm. *Fruit*: Pedicel 2-3.5(-6.5) cm long. Fruit 2.5-4.5 cm long; 1.2-2.5 cm in diam; aculeate; glabrous. Aculei (4-)6-8(-12) mm long; 0.7-1.4(-2) mm in diam. *Seeds*: 4.3-5 X 2-2.2 mm.

Distribution: Africa: Chad, Egypt, Ethiopia, Kenya, Mauritania, Niger, Rwanda, Somalia, Tanzania, and Uganda. Southwest Asia: Saudi Arabia, South Yemen, and Yemen.

Cucumis prophetarum was traditionally recognized as a species of dry habitats, principally deserts, ranging from Senegal across drier areas of northern Africa with Kenya as its southern limit in eastern Africa, then northwards to Syria, throughout the Arabian peninsula, and finally eastwards into Pakistan and drier areas of northwestern India. Jeffrey (1962) broadened the concept of *C. prophetarum* and included in it an element from tropical east Africa ranging from Tanzania and Uganda to Egypt, Socotra, and the Arabian peninsula for which he made the new combination *C. prophetarum* subsp. *dissectus*. Jeffrey (1967; ms. for "Flora of Ethiopia") distinguished the subspecies by gray-green coloration of plants, rather dense appressed hairs, and fruit spines up to 3 mm long in subsp. *prophetarum* and bright green coloration of plants, rather slender more or less spreading hairs, and fruit spines 4-6 mm long in subsp. *dissectus*. Collectors (C. Jeffrey, pers. comm.; M. Thulin, mss.) have reported that subsp. *prophetarum* occurs in dry habitats and that subsp. *dissectus* is found in more mesic habitats. Differentiation within the species is apparently correlated with ecological adaptation.

Plant coloration, the character most frequently and consistently used to separate these two subspecies, can be difficult to interpret, and length of the aculei on the fruits overlaps, those of subsp. *prophetarum* being aculei 1-7 mm and those of subsp. *dissectus* (4-)6-8(-12) mm long. Morphological structures other than fruit and seed are generally smaller in subsp. *prophetarum* than in subsp. *dissectus*; for measurements see the key to the subspecies. In all structures there is overlap in the size ranges, except seed size and female-flower pedicel length.

The surface of the fruits of some specimens of subsp. *prophetarum* are both aculate and hispidulous. Individual specimens of this sort have been collected in Egypt, Ethiopia, Somalia, Israel, Saudi Arabia, and Pakistan, and they range in habit from very woody to less so and in leaf morphology from deeply lobed to almost entire. The condition must have arisen more than once because the individuals exhibiting it have varying habits and leaf types and occur in widely separated areas. Thulin (1991) has delimited *C. pubituberculatus* as consisting of collections from Somalia with both aculei and pubescence on the fruit, deeply narrowly lobed leaves, petioles with 2 types of pubescence, larger male flowers, and short aculei on the fruits. If similar criteria were applied across all of *C. prophetarum* throughout its range, two or possibly three additional new species would have to be recognized. Biosystematic studies are needed to understand the genetics of pubescence and aculei appearing together on fruits so that the circumscription of *C. prophetarum* can be established.

No material referrable to the epithet *C. halabarda* was available to me for study. According to Thulin (1991), this rare taxon is endemic to Somalia. Thulin (1991) considered four collections to be members of the species, the two syntypes and two recent collections, *Thulin & Abdir Dahir 6443* and *Gillett, Hemming, & Watson 22225*. He believed that it should be maintained as a distinct species because its male-flower petals are longer and narrower than those of *C. prophetarum* and also acute at the apex, and Jeffrey (pers. comm.) has agreed with him. Thulin (pers. comm.) will be presenting it as a distinct species in his forth coming treatment of *Cucumis* for the *Flora of Somalia*. The available data does not, in my opinion, justify its maintenance as a distinct species, but further studies are needed to clarify its status.

11. **Cucumis pubituberculatus** Thulin, Nord. J. Bot. 11(5): 538. 1991.—TYPES: SOMALIA. GALGUDUD REGION: Ceeldheer, just E of town, 3°51'N, 47°10'E, 7 May 1990, *M. Thulin, Hedrén & Abdi Dahir 7288* (holotype: UPS; isotypes: E, FT, K, MOG). SHABEELLAHA DHEXE REGION: Ca 11 km NE of Muqdisho on road to Warshiikh, 16 May 1989, *M. Thulin 6321* (paratypes: K, UPS).

Plants: Herbs; perennial; with a woody rootstock; monoecious. Stems procumbent; sulcate (older parts whitish, younger green); not aculeate; antrorsely scabrous. Nodes not geniculate. Internodes 1.5-4 cm long. *Leaves*: Petioles 0.3-1 cm long; not aculeate; pubescence 2 different types uniformly intermixed on each petiole. Leaf blades 3-palmately lobed; with the margin serrate; ovate in outline; truncate to subcordate at the base; with or without a basal sinus; acute to obtuse at the apex; 1.5-5 X 0.8-3 cm; scabrous on the upper surface (hairs slightly longer and more curved); not aculeate on the veins below; scabrous on the veins below; scabrous on the intervenium below. Central leaf-blade lobe symmetrical; entire or pinnatifid; narrowly oblong in outline; acute to obtuse at the apex; 1.5-4.5 X 0.5-1.4 cm. Lateral leaf-blade lobes symmetrical or asymmetrical; entire or pinnatifid; elliptic in outline; acute to obtuse at the apex; 0.4-1.2 X 0.3-0.7 cm. *Tendrils*: Present; solitary; simple; 1-2 cm long; not aculeate; scabrous. *Inflorescences*: Unisexual. *Male inflorescences*: A solitary flower or fasciculate; 1- or 2-flowered; sessile. *Male flowers*: Pedicellate. Pedicel 8-15 mm long; scabrous; without bracteoles. Hypanthium infundibular; 5-6.5 mm long; scabrous. Calyx lobes linear in outline (with a broad base); 1.5-2.5 mm long; scabrous. Corolla infundibular; scabrous outside. Corolla lobes elliptic to oblong in outline; obtuse at the apex; 8-12 mm long; scabrous outside. Stamens separating from the hypanthium 2-3 mm from the base of the hypanthium. Anther thecae puberulent. Anther connective obovate; unilobate; obtuse at apex; papillate to fimbriate at the apex; 0.8-1.6 mm long. Disc cylindrical. *Female inflorescences*: A solitary flower; sessile. *Female flowers*: Pedicellate. Pedicel 15-45 mm long; scabrous; cylindrical. The lower 2/3s to 3/4s of hypanthium fused to the ovary. Fused portion of hypanthium ellipsoid; 11-15 mm long; 6-8 mm in diam; aculeate; puberulent. The upper 1/3 to 1/4 of hypanthium free from the ovary. Free portion of hypanthium campanulate; ca 4 mm long; not aculeate outside; hispid and puberulent outside. Calyx lobes linear in outline (with a broad base); narrowly acute at the apex; 1.2-1.6 mm long; scabrous. Corolla tube present. Corolla lobes obovate to oblong in outline; obtuse at the apex; ca 9.5 mm long. Staminodes present; glabrous. Style ca 1.6 mm long. Stigma globose; ca 2.8 mm long; ca 3.2 mm in diam; entire. *Fruit*: Not geocarpic, maturing above ground and readily visible. Pedicel cylindrical. Fruit green (when immature); weakly globose; ca 2.5 cm long; ca 2.5 cm in diam; sparsely aculeate; puberulent; blunt at the apex. Aculei terete; ca 1 mm long. *Seeds*: Elliptic (acute at one end); ca 3.5 X ca 2 mm; unwinged.

Distribution: Africa: central, coastal Somalia.

In the field, *C. pubituberculatus* has a distinct facies (Thulin, pers. comm.). Thulin (pers. comm.) has found it growing together with *C. prophetarum* subsp. *dissectus*, and *C. prophetarum* subsp. *prophetarum* was also found in the same area. Uniform application of Thulin's criteria to *C. prophetarum* could lead to the splintering off from it of various microspecies. Biosystematics studies are needed to clarify the circumscription of *C. prophetarum* and its relationship to *C. pubituberculatus*.

12. **Cucumis zeyheri** Sonder in Harvey & Sonder, Fl. cap. 2: 496. 1862.—Types: SOUTH AFRICA. Cape Province: Gamka River, May 1840, *C. Zeyher 582* (lectotype: BM!; isolectotypes: K! MEL! S!).—Natal: Port Natal [Durban], *W. Gueinzius 398* (paratypes: LE! S! W!).—Transvaal: Magaliesberg, Nov 1840, *C. Zeyher 583* (syntype: BM!).—*Cucumis africanus* Linnaeus filius var. *zeyheri* (Sonder in Harvey & Sonder) J.B. Davy, Man. pl. Transvaal 39. 1926.—*Cucumis prophetarum* Linnaeus subsp. *zeyheri* (Sonder in Harvey & Sonder) C. Jeffrey, Kew Bull. 15(3): 35. 1962.

 Cucumis diniae van Raamsdonk & Visser, Nord. J. Bot. 12(3): 333. 1992.—Types: CULTIVATED. Netherlands: greenhouse Wageningen, *T. de Vries 1007-1012, 1103*, accession Gbn (CPRO-DLO genebank number) *1053* plant 2, seedlings of collection made in Cape Province, South Africa (holotype: WAG (*1011*); isotypes: GAT K LE NY WAHO), *1001-1005* and *1013-1025*, accession Gbn (CPRO-DLO genebank number) *1053* plant *1, 3, 4,* and *5*, seedlings of collection made in Cape Province, South Africa (paratypes: GAT K LE NY WAG WAHO). SOUTH AFRICA. Cape Province: Hay Div., Griquatown, Feb 1937, *M. Wilman PRE42851* (paratype: PRE); District of Jansenville, 1.5 miles SW of Jansenville, 457 m, 26 Mar 1957, *J.P.H. Acocks 19218* (paratypes: K! M! SRGH).

Plants: Herbs; perennial; with a woody rootstock; lacking tubers; monoecious. Stems procumbent; sulcate; not aculeate; retrorse-strigose or antrorse-strigose (rarely); with nonbreakaway hairs. Stem hairs 0.4-0.7(-1) mm long. Nodes not geniculate. Internodes 3-5.5 cm long. *Leaves*: Petioles 1.5-4 cm long; sulcate; not aculeate; pubescence a single type on each petiole; retrorse-strigose or antrorse-strigose (rarely); with nonbreakaway hairs. Leaf blades 5-palmately lobed; with the margin serrate; ovate to broadly ovate in outline; cordate at the base; with a basal sinus; 0.4-0.8 cm deep; acute to narrowly acute at the apex; 3-6.5 X 2.2-5.5 cm; ca 1.2 times longer than wide; hispidulous to scabrous on the upper surface; not aculeate on the veins below; hispidulous to antrorse-strigose on the veins below; hispidulous to scabrous on the intervenium below; with nonbreakaway hairs. Central leaf-blade lobe pinnatifid; elliptic to narrowly elliptic in outline; acute to narrowly acute at the apex; 2.3-4.6 X 1-2.6 cm. Lateral leaf-blade lobes asymmetrical; entire; broadly elliptic to elliptic in outline; broadly acute at the apex; 0.6-2.6 X 0.4-2.1 cm. *Tendrils*: Present; solitary; simple; 2-2.5(-5) cm long; not aculeate; sparsely hispidulous (on the basal half). *Inflorescences*: Unisexual. *Male inflorescences*: A solitary flower or fasciculate (rarely); 1- or 2-flowered; sessile. *Male flowers*: Pedicel terete in cross section; 8-28 mm long; hispidulous; without bracteoles. Hypanthium campanulate; 3.2-5.3 mm long; 2.7-2.9 mm in diam; hispidulous. Calyx lobes narrowly oblong in outline; narrowly acute at the apex; 1.3-3.8 X 0.3-0.6 mm; hispidulous (to sparsely so). Corolla infundibular; sparsely hispidulous outside; sparsely puberulent inside. Corolla tube 0.6-1.2 mm long; 3.6-4 mm in diam; sparsely hispidulous outside. Corolla lobes elliptic or obovate in outline; acute, or broadly acute and mucronate at the apex; 4.8-6.8 X 2.8-4 mm; sparsely hispidulous outside; sparsely puberulent inside. Stamens separating from the hypanthium 1.3-1.6 mm from the base of the hypanthium. Filaments terete in cross section; 0.6-0.8 mm long; 0.3-0.5 mm wide;

glabrous. Anther thecae 2-2.4 mm long; puberulent. Anther connective obovate; unilobate; obtuse at apex; papillate at the apex; 1.2-1.6 X ca 0.8 mm. Disc cylindrical; 0.6-0.9 mm long; 0.8-1.3 mm in diam. *Female inflorescences*: A solitary flower. *Female flowers*: Pedicel terete in outline; 6-15 mm long; hispidulous; with nonbreakaway hairs; cylindrical. The lower 2/3 of hypanthium fused to the ovary. Fused portion of hypanthium ellipsoid; 4-8.8 mm long; 3.2-4.8 mm in diam; aculeate; glabrous. Aculei soft; 1.6-2.4 mm long. Hyaline bristle at apex of aculeus 0.7-1.2 mm long; 0.6-3 times as long as the opaque base. The upper 1/3 of hypanthium free from the ovary. Free portion of hypanthium 3.2-4 mm long; 2.4-2.7 mm in diam; not aculeate outside; sparsely hispidulous outside; sparsely puberulent inside. Calyx lobes linear to narrowly oblong in outline; narrowly acute at the apex; 1.9-3.7 X 0.3-0.6 mm; sparsely hispidulous. Corolla sparsely hispidulous outside; sparsely puberulent inside. Corolla tube present; 0.8-2 mm long; 2.4-3.2 mm in diam; sparsely hispidulous outside. Corolla lobes elliptic to broadly elliptic in outline; obtuse and mucronate at the apex; 4-4.8 X 2.4-3.6 mm; sparsely hispidulous outside; puberulent inside. Staminodes present; separating from the free portion of the hypanthium ca 1.6 mm above the ovary; ca 0.6 mm long; 0.3-0.5 mm in diam; glabrous. Style 0.8-1.6 mm long; 0.4-0.8 mm in diam; subtended by a circular disc. Disc 0.4-0.8 mm long; 0.9-1.3 mm in diam. Stigma 1.4-2.6 mm long; 1.6-2.4 mm in diam; lobate; with 6 finger-like projections on the margin. Finger-like stigmatic projections 1-2 mm long; 0.6-0.8 mm in diam; papillate. *Fruit*: Not geocarpic, maturing above ground and readily visible. Pedicel terete in cross section; 1.5-2(-4) cm long; sparsely hispidulous; with nonbreakaway hairs; cylindrical. Fruit monocolored (when mature) or bicolored with longitudinal stripes from base to apex (when immature); dark green (when immature) or yellow (when mature); with light green longitudinal stripes (when immature); ellipsoid; 4-5 cm long; 2.5-4 cm in diam; aculeate; glabrous; blunt at the apex. Aculei terete; with hyaline bristle lost; 1.4-7.2 or 9-13 mm long; 0.8-1.1 mm in diam. *Seeds*: Elliptic; 5.6-6.7 X 2.8-3.1 mm; 1.4-1.6 mm thick; unwinged. *Chromosome number*: 2n = 24 (Deakin, Bohn & Whitaker, 1971; Dane & Tsuchiya, 1976; Varekamp, Visser & Nijs, 1982; Ramachandran, 1984; Singh & Yadava, 1984; Nijs & Visser, 1985; Ramachandran & Narayan, 1985) or 48 (Dane & Tsuchiya, 1976, 1979; Varekamp, Visser & Nijs, 1982; Ramachandran, 1984; Nijs & Visser, 1985; Ramachandran & Narayan, 1985).

Distribution: Africa: Lesotho, Mozambique, South Africa (Cape Province, Natal, and Transvaal), Swaziland, Zambia, and Zimbabwe.

When publishing *C. zeyheri*, Sonder (1862) cited *Zeyher 582* and *583* of his herbarium under the habitat of the species and in the discussion referred to *Gueinzius 398* as a similar plant lacking fruit. *Zeyher 582* and *583* are syntypes, and *Gueinzius 398* is a paratype. *Zeyher 582* at BM is here selected as lectotype, therefore the duplicates of *Zeyher 583* are syntypes.

Based on morphological criteria, Jeffrey (1962) reduced *C zeyheri* to a subspecies of *C. prophetarum*. Deakin et al. (1971) hybridized *C. prophetarum*, s.s. and *C. zeyheri*. The cross yielded vigorous F_1 hybrids, but they were weakly fertile. Male flowers of the hybrid had an average pollen stainability of 37%. They recommended that *C. zeyheri* be maintained as a species, and Jeffrey (1975) agreed.

Van Raamsdonk and Visser (1992) established *C. diniae* using greenhouse grown material which was not available for my study. In the protologue, they also cited two wild collections, *M. Wilman PRE42851* and *J.P.H. Acocks 19218*, from South Africa which are, therefore, paratypes. Duplicates of *Acocks 19218* do not exhibit the leaf characters stressed by van Raamsdonk and Visser (1992) and fit well within the morphological variation of *C. zeyheri* as set forth above.

The consistent morphological differences between *C. zeyheri* and *C. prophetarum* are set out in couplet 25 of the Key to the Species of *Cucumis*. The most easily observed differences are in the size of their seeds, 4.3-6 mm long, 2-2.6 mm wide, and 1-1.3 mm thick in *C. prophetarum* and 5.6-6.7 mm long, 2.8-3.1 mm wide, and 1.4-1.6 mm thick in *C. zeyheri*. They are also allopatric; *C. zeyheri* is found only in southern Africa (Lesotho, Mozambique, South Africa, Swaziland, Zambia, and Zimbabwe) and *C. prophetarum* in tropical east Africa, northeastern Africa, southwest Asia, northeastern India and Pakistan. The two species are maintained here on the basis of their genetic incompatibility and allopatry.

13. **Cucumis prolatior** Kirkbride, sp. nov.—TYPES: KENYA. Eastern: District of Kitui, Nairobi-Garissa Road, 3 km SW of Mwingi, 0°56'S, 38°2'E, 960 m, 8 May 1974, *J.B. Gillett & F.N. Gachathi 20478* (holotype: K!).—Rift Valley: Ngare Narok, [1°13'N, 36°33'E - 0°52'N, 36°47'E (Polhill, 1988)], 1524 m, 26 Apr 1944, *P.R.O. Bally B3645* (paratype: K!). Fig. 2.

A *Cucumere prophetarum* folii lamina basi acuta, folii lobis apice anguste acutis, corollae lobis apice anguste acutis, fructuumque aculeis 20-22 mm longis et 2.4-3.2 mm diametro absimilis.

Plants: Herbs; monoecious. Stems procumbent; sulcate; not aculeate; hirsute; with nonbreakaway hairs. Stem hairs 0.6-1.4 mm long. Nodes not geniculate. Internodes 3.5-9.5 cm long. *Leaves*: Petioles 2-3.5 cm long; sulcate; not aculeate; pubescence a single type on each petiole; hirsute; with nonbreakaway hairs. Leaf blades 5-palmately lobed; with the margin serrate; ovate in outline; acute at the base; without a basal sinus; narrowly acute at the apex; 5-6 X 4-4.5 cm; 1.2-1.3 times longer than wide; hirsute on the upper surface; not aculeate on the veins below; hirsute on the veins below; hirsute on the intervenium below; with nonbreakaway hairs. Central leaf-blade lobe entire; narrowly elliptic in outline; narrowly acute at the apex; 4-5 X 0.7-1 cm. Lateral leaf-blade lobes symmetrical or asymmetrical; entire; narrowly elliptic in outline; narrowly acute at the apex; 1.5-3.5 X 0.5-0.7 cm. *Tendrils*: Present; solitary; simple; 2-5 cm long; not aculeate; hirsute. *Inflorescences*: Unisexual. *Male inflorescences*: A solitary flower or fasciculate; 1- or 2-flowered; sessile. *Male flowers*: Pedicel terete in cross section; 10-15 mm long; hirsute; without bracteoles. Hypanthium campanulate; 4-4.4 mm long; ca 2.4 mm in diam; hirsute. Calyx lobes linear in outline; narrowly acute at the apex; 0.8-1.2 X ca 0.2 mm; hispidulous. Corolla infundibular; sparsely hirsute outside; glabrous inside. Corolla tube ca 0.8 mm long; ca 3.2 mm in diam; sparsely hirsute outside. Corolla lobes elliptic in outline; narrowly acute at the apex; ca 6 X ca 2.8 mm; sparsely hirsute outside; glabrous inside. Stamens separating from the hypanthium ca 2 mm from the base of the hypanthium. Filaments radially compressed in cross section; ca 0.7 mm long; ca 0.3 mm wide; glabrous. Anther thecae ca 2.2 mm long; puberulent. Anther connective obovate; unilobate; obtuse at apex; papillate at the apex; ca 1.6 X 0.7-0.8 mm. Disc cylindrical; ca 1 mm long; ca 1.7 mm in diam. *Fruit*: Not geocarpic, maturing above ground and readily visible. Pedicel terete in cross section; 5.5-10.5 cm long; sparsely hirsute (most hairs broken off); with nonbreakaway hairs; cylindrical. Fruit bicolored with longitudinal stripes from base to apex (when immature); green; with purple longitudinal stripes; ellipsoid; 5-6 cm long; ca 3 cm in diam; aculeate; glabrous; blunt at the apex. Aculei terete; with hyaline bristle lost; 20-22 mm long; 2.4-3.2 mm in diam. *Seeds*: Elliptic; 4.4-4.8 X 2.4-2.6 mm; 0.8-1 mm thick; unwinged.

Distribution: Africa: central and south central Kenya.

Cucumis prolatior is known from only two specimens. Jeffrey (1967) recognized *Bally B3645* as *Cucumis* sp. A in the *Flora of Tropical East Africa*. *Bally B3645* is an older specimen, now leafless with a single small male bud and two mature fruits. *Gillett & Gachathi 20478* is a younger specimen with intact leaves, male flowers in various stages of development, and a single immature fruit. Neither specimen has female flowers. Consequently, the vegetative

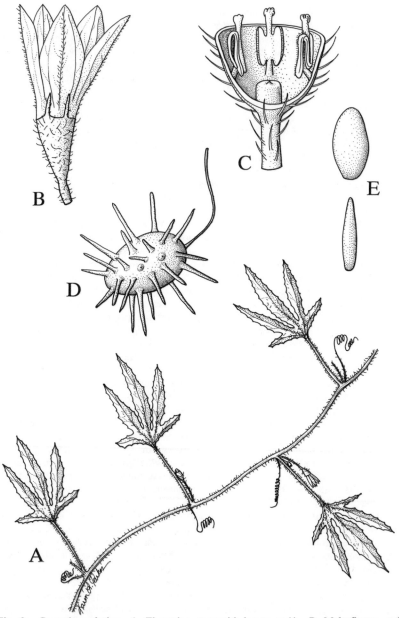

Fig. 2. *Cucumis prolatior*. A. Flowering stem with leaves, x ½. B. Male flower, x 5. C. Male flower, opened to show stamens and disc, x 7½. D. Fruit, x ½. E. Seed, face and side views, x 5. (*Bally B3645* (K), D, E; *Gillett & Gachathi 20478* (K), A, B, C.)

and male flower descriptions are of *Gillett & Gachathi*, and the fruit and seed descriptions are from *Bally*.

Cucumis prolatior is very similar to *C. prophetarum*, especially the variations encompassed by subsp. *dissectus* (Naudin) C. Jeffrey, but it is easily distinguished from it and all other species of *Cucumis* by fruit aculei more than twice as long as any known elsewhere in the genus, the character that suggested the epithet.

14. **Cucumis insignis** C. Jeffrey, Kew Bull. 40(1): 209, *f. 1.* 1985.—TYPE: ETHIOPIA. Sidamo: 39 km from Yavello on new road to Agree Mariam, ca 1 km S of village of Sorupa, 5°14'N, 38°12'E, 1580 m, *Gilbert & Jefford 4656* (holotype: K).

Plants: Herbs; perennial; monoecious. Stems procumbent (grayish); not aculeate; hispidulous. Nodes not geniculate. Internodes 5.5-9 cm long. *Leaves*: Petioles 0.7-3.8 cm long; not aculeate; pubescence a single type on each petiole; hispid. Leaf blades 5- or 3-palmately (rarely) lobed; with the margin entire; broadly ovate in outline; cordate at the base; with a basal sinus; 0.4-0.8 cm deep; acute at the apex; 2.8-9 X 1.5-7.8 cm; 1.1-1.9 times longer than wide; hispid (grayish) on the upper surface; hispid (grayish) on the intervenium below. Central leaf-blade lobe pinnatifid; narrowly elliptic in outline; narrowly acute at the apex; 5.7-7.4 X 1.4-2.1 cm. Lateral leaf-blade lobes asymmetrical; pinnatifid; elliptic to narrowly elliptic in outline; obtuse or acute at the apex; 1.4-4.5 X 0.8-1.2 cm. *Tendrils*: Present; solitary; simple or bifid (both types occurring on an individual plant); 1-3 cm long. *Inflorescences*: Unisexual. *Male inflorescences*: Fasciculate; 3-flowered; sessile. *Male flowers*: Pedicel ca 30 mm long; without bracteoles. Hypanthium campanulate; ca 4 mm long; ca 3 mm in diam; hispid. Calyx lobes narrowly triangular in outline; narrowly acute at the apex; ca 2 X ca 0.5 mm; hispid. Corolla infundibular. Corolla lobes ovate (towards oblong) in outline; acute at the apex; ca 6 X ca 2.5 mm. *Female inflorescences*: A solitary flower. *Female flowers*: Pedicel 75-100 mm long; weakly flaring upwards from a narrower base to a wider apex. The lower 3/4 of hypanthium fused to the ovary. Fused portion of hypanthium ellipsoid; ca 15 mm long; ca 5 mm in diam; aculeate; glabrous. Aculei stiff; 3-4 mm long. Hyaline bristle at apex of aculeus ca 1 mm long; 0.3-0.5 times as long as the opaque base. The upper 1/4 of hypanthium free from the ovary. Free portion of hypanthium ca 3.5 mm long; ca 3 mm in diam; not aculeate outside; sparsely hispid outside. Calyx lobes narrowly triangular in outline; narrowly acute at the apex; 1.5-2.5 X ca 0.5 mm; sparsely hispidulous. Corolla lobes elliptic in outline; acute at the apex; ca 5 X ca 2 mm. *Fruit*: Not geocarpic, maturing above ground and readily visible. Pedicel 10.5-12 cm long; weakly flaring upwards from a narrower base to a wider apex. Fruit dark green; with light green longitudinal stripes; ellipsoid; 4-4.5 cm long; ca 3.2 cm in diam; aculeate; glabrous; blunt at the apex. Aculei 10-16 mm long. *Seeds*: Elliptic; 5-5.5 X 2.3-2.5 mm; ca 1 mm thick; unwinged.

Distribution: Africa: Ethiopia.

Cucumis insignis is the only member of the genus *Cucumis* to have both simple and bifid tendrils on the same individual plant, and thereby is distinguished from all other species of the genus. It is known only from the type collection, which I have not seen, the above description being extracted from the protologue.

15. **Cucumis globosus** C. Jeffrey, Kew Bull. 19(2): 215. 1965.—TYPES. TANZANIA. Mbeya District: About 1 km southeast of Ilomba local court, 27 Feb 1963, *A. Harwood 19* (holotype: K); Isyesye, about 1.6 km southeast of Ilomba local court, 3/10/1964, *A. Harwood 73* (paratypes: K! LISC! SRGH!), 3/1964, *A. Harwood 92* (paratype: K!).

Plants: Herbs; perennial; with a woody rootstock; lacking tubers; monoecious. Stems procumbent; sulcate; not aculeate; retrorse-strigose; with nonbreakaway hairs. Stem hairs 0.4-0.5 mm long. Nodes not geniculate. Internodes 2-4 cm long. *Leaves*: Petioles 0.5-1(-1.5) cm long; sulcate; not aculeate; pubescence a single type on each petiole; retrorse-strigose; with nonbreakaway hairs. Leaf blades 3- or 5-palmately lobed; with the margin serrate; ovate in outline; subtruncate to broadly acute, or cordate at the base; with or without a basal sinus; 0.3-0.5 cm deep; acute to broadly acute at the apex; 5.5-8 X 2.5-4.5 cm; 1.5-2 times longer than wide; scabrous on the upper surface; not aculeate on the veins below; scabrous on the veins below; scabrous on the intervenium below; with nonbreakaway hairs. Central leaf-blade lobe entire; narrowly elliptic to narrowly ovate in outline; acute to broadly acute at the apex; 4-6.5 X 1-2 cm. Lateral leaf-blade lobes asymmetrical; entire; ovate or narrowly elliptic in outline; acute to broadly acute at the apex; 0.5-3 X 0.3-0.8 cm. *Tendrils*: Present; solitary; simple; ca 1.5 cm long; not aculeate; scabrous. *Inflorescences*: Unisexual. *Male inflorescences*: A solitary flower; sessile. *Male flowers*: Pedicel terete in cross section; 5-8 mm long (fide Jeffrey, 1965, 1967); scabrous; without bracteoles. Hypanthium infundibular; 3.2-4 mm long; 2-2.5 mm in diam; scabrous. Calyx lobes linear in outline; narrowly acute at the apex; 1.8-3 X ca 0.2 mm; scabrous. Corolla infundibular; apically hispidulous and basally puberulent outside. Corolla tube ca 0.4 mm long; ca 2.4 mm in diam; puberulent outside. Corolla lobes ca 8 X ca 4 mm (fide Jeffrey, 1965, 1967); hispidulous outside. *Female inflorescences*: A solitary flower. *Female flowers*: Pedicel sulcate in outline; 4.8-5.6 mm long; hispidulous; with nonbreakaway hairs; cylindrical. The lower 2/3 of hypanthium fused to the ovary. Fused portion of hypanthium ellipsoid; 5.6-6 mm long; 2.8-3.2 mm in diam; aculeate; glabrous. Aculei soft; 0.8-1 mm long (the opaque basal portion short). Hyaline bristle at apex of aculeus 0.6-0.8 mm long; 3-4 times as long as the opaque base. The upper 1/3 of hypanthium free from the ovary. Free portion of hypanthium 2.4-2.8 mm long; 2.4-2.8 mm in diam; not aculeate outside; hispidulous outside. Calyx lobes linear in outline; narrowly acute at the apex; 1.8-3.6 X 0.3-0.4 mm; scabrous. Corolla hispidulous outside. Corolla tube present; ca 0.4 mm long; ca 2.4 mm in diam; hispidulous outside. Corolla lobes hispidulous outside. *Fruit*: Not geocarpic, maturing above ground and readily visible. Pedicel cylindrical (fide Jeffrey, 1965). Fruit bicolored with longitudinal stripes only on upper portion (fide Jeffrey, 1965, 1967); light green (fide Jeffrey, 1967); with dark green longitudinal stripes (fide Jeffrey, 1967); globose (fide Jeffrey, 1965, 1967); 4-8 cm long (fide Jeffrey, 1965, 1967); 4-8 cm in diam (fide Jeffrey, 1965, 1967); sparsely aculeate (with the aculei short, fide Jeffrey, 1965, 1967); glabrous; blunt at the apex. *Seeds*: Elliptic (fide Jeffrey, 1965, 1967); ca 8.5 X ca 5.5 mm (fide Jeffrey, 1965, 1967); ca 3.7 mm thick (fide Jeffrey, 1965, 1967); unwinged (fide Jeffrey, 1965).

Distribution: Africa: Tanzania.

Cucumis globosus is a rare, localized endemic known only from the holotype and paratype collections.

16. **Cucumis thulinianus** Kirkbride, sp. nov.—TYPES: SOMALIA. Northern region, Erigavo, 1722 m, 5 Oct 1960, *C.F. Hemming 2011* (holotype: K!); Erigavo Grazing Reserve, near guest-house, 18 Jan 1973, *P.R.O. Bally & R. Melville 16028* (paratype: K!). Fig.3.

A *Cucumere prophetarum* foliorum lamina basi sagittata, floris feminei pedicello circa 6 mm longo, fructuum aculeis 10-15 mm altis et 3-5 mm diametro differt.

Plants: Herbs. Stems procumbent or climbing; sulcate; not aculeate; hispidulous pubescence in the grooves and retrorse-strigose on the ridges between the grooves; with nonbreakaway hairs. Stem hairs 0.2-0.3 (hispidulous pubescence) or 0.5-1 (retrorse-strigose

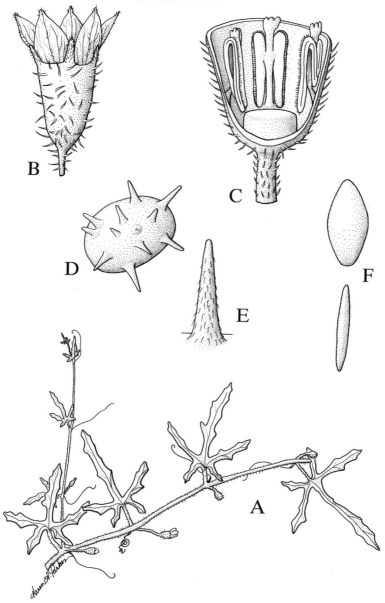

Fig. 3. *Cucumis thulinianus*. A. Flowering stem with leaves, x 3/4. B. Male flower, x 7. C. Male flower, opened to show stamens and disc, x 11. D. Fruit, x ½. E. Fruit aculeus with retrorse strigose pubescence, x 2. F. Seed, face and side views, x 5. (*Bally & Melville 16028* (K), D, E, F; *Hemming 2011* (K), A, B, C.)

pubescence) mm long. Nodes not geniculate. Internodes 2-4 cm long. *Leaves*: Petioles 0.5-1 cm long; terete to sulcate in cross section; not aculeate; pubescence 2 different types uniformly intermixed on each petiole; hispidulous and retrorse-strigose; with nonbreakaway hairs. Leaf blades 5-palmately lobed; with the margin serrate; ovate to broadly ovate in outline; sagittate at the base; with a basal sinus; 0.3-0.7 cm deep; acute at the apex; 2.5-4 X 1.3-3.5 cm; 1.1-1.9 times longer than wide; antrorse-strigose on the upper surface; not aculeate on the veins below; hispidulous or antrorse-strigose on the veins below; hispidulous or antrorse-strigose on the intervenium below; with nonbreakaway hairs. Central leaf-blade lobe pinnatifid; narrowly elliptic in outline; acute at the apex; 2-3.5 X 0.6-0.8 cm. Lateral leaf-blade lobes symmetrical or asymmetrical; entire or pinnatifid; narrowly elliptic in outline; narrowly acute to acute at the apex; 0.5-2 X 0.2-0.6 cm. *Tendrils*: Present; solitary; simple; 3.5-5 cm long; not aculeate; hispidulous or antrorse-strigose (apically sparser than basally). *Inflorescences*: Unisexual. *Male inflorescences*: A solitary flower; sessile. *Male flowers*: Pedicel terete in cross section; 8-10 mm long; hispidulous; without bracteoles. Hypanthium campanulate; 3.6-4.4 mm long; 2.4-3 mm in diam; hispidulous. Calyx lobes narrowly triangular in outline; narrowly acute at the apex; ca 1.4 X 0.4-0.5 mm; short hispidulous. Corolla infundibular; hispidulous on the veins outside; puberulent inside. Corolla tube 0.6-1 mm long; ca 3.2 mm in diam; hispidulous on the veins outside. Corolla lobes broadly elliptic in outline; acute and mucronate at the apex (the mucro ca 0.6 mm long and short hispidulous); 3.6-4.4 X 2.7-3.2 mm; hispidulous on the veins outside; puberulent inside. Stamens separating from the hypanthium ca 1.6 mm from the base of the hypanthium. Filaments terete in cross section; ca 0.4 mm long; 0.5-0.6 mm wide; glabrous. Anther thecae 2.7-2.9 mm long; puberulent. Anther connective obovate; unilobate; obtuse at apex; papillate at the apex; 1-1.4 X 0.6-1 mm. Disc cylindrical; ca 1 mm long; ca 1.8 mm in diam. *Female inflorescences*: A solitary flower. *Female flowers*: Pedicel terete in outline; ca 6 mm long; hispidulous; with nonbreakaway hairs; cylindrical. The lower 2/3 of hypanthium fused to the ovary. Fused portion of hypanthium ellipsoid (in bud); ca 2.6 mm long; ca 1 mm in diam; aculeate; glabrous. Aculei 0.6-0.9 mm long (in bud). Hyaline bristle at apex of aculeus 0.4-0.6 mm long (in bud); ca 2 times as long as the opaque base. The upper 1/3 of hypanthium free from the ovary (in bud). Free portion of hypanthium ca 1 mm long; ca 0.8 mm in diam; not aculeate outside; hispidulous outside. *Fruit*: Not geocarpic, maturing above ground and readily visible. Fruit monocolored; yellow; ellipsoid; ca 5 cm long; ca 3.5 cm in diam; aculeate; retrorse-strigose; blunt at the apex. Aculei terete; with hyaline bristle lost; 10-15 mm long; 3-5 mm in diam. *Seeds*: Elliptic; 5.2-5.6 X ca 2.6 mm; ca 0.8 mm thick; unwinged.

Distribution: Africa: Somalia, only near Erigavo.

In Thulin's *Cucumis* manuscript for the *Flora of Somalia*, this species is treated as number 10, *Cucumis* sp. He is correct in maintaining this as a separate species, so I am naming it in his honor to recognize his excellent work on the Cucurbitaceae for the *Flora*. It is most similar to *C. prophetarum* but easily separated by the characters set forth in couplet 31 of the Key to the Species of *Cucumis*.

17. **Cucumis ficifolius** A. Richard, Tent. fl. abyss. 1: 294, *t. 53 bis.* 1847. *Cucumis figarei* Delile ex Naudin, nom. illeg., Ann. Sci. Nat. Bot., sér. 4, 11: 16. 1859.—TYPES: ETHIOPIA. Sholoda, *R. Quartin-Dillon & A. Petit s.n.* (lectotype: P! [Kew negative 4341]; isolectotype: P!); *R. Quartin-Dillon & A. Petit s.n.* (syntype: P!) [designated by Jeffrey, 1967].

   Bryonia obtusa A. Richard, Tent. fl. abyss. 1: 289. 1847.—TYPES: ETHIOPIA. Choho, *A. Petit s.n.* (holotype: P!; isotype: P!).

Cucumis abyssinicus A. Richard, Tent. fl. abyss. 1: 294. 1847.—TYPES: ETHIOPIA.
Chiré, *R. Quartin-Dillon & A. Petit s.n.* (holotype: P! [Kew negative 4339];
isotype: P!).
Cucumis figarei Delile ex Naudin var. *cyrtopodus* Naudin, Ann. Sci. Nat. Bot., sér. 4,
11: 17. 1859.—TYPES: ETHIOPIA. Sholoda, *R. Quartin-Dillon & A. Petit
s.n.* (lectotype: P! [Kew negative 4340]); Kiéa, Quartin- *R. Dillon & A. Petit
s.n.* (syntype: P!); Adua, *R. Quartin-Dillon & A. Petit s.n.* (syntype: P!); *R.
Quartin-Dillon & A. Petit s.n.* (syntype: P!) [Jeffrey, 1967].
Cucumis figarei Delile ex Naudin var. *microphyllus* Naudin, Ann. Sci. Nat. Bot., sér. 4,
11: 17. 1859.—TYPE: ETHIOPIA. *C.L.X. Rochet d'Héricourt s.n.* (holotype:
P! [Kew negative 4343]).

Plants: Herbs; perennial; with a woody rootstock; lacking tubers; monoecious. Stems
procumbent; sulcate; aculeate (aculei weakly retrorse, 1-1.6 mm long); hispidulous; with
nonbreakaway hairs. Stem hairs 0.3-0.6 mm long. Nodes not geniculate. Internodes 1.5-6(-12)
cm long (weedy specimens from cultivated areas exhibit extreme values). *Leaves*: Petioles
1-2.5(-6) cm long (weedy specimens from cultivated areas exhibit extreme values); sulcate;
aculeate (aculei weakly retrorse); pubescence a single type on each petiole; hispidulous; with
nonbreakaway hairs. Leaf blades 5-palmately lobed; with the margin serrate; ovate to broadly
ovate in outline; subcordate to cordate at the base; with a basal sinus; 0.5-1 cm deep; broadly
acute to obtuse at the apex; (1.5-)3.5-6(-9.5) X (1.5-)3-5(-9.5) cm (weedy specimens from
cultivated areas exhibit large, extreme values); 1-1.1 times longer than wide; antrorse-strigose
on the upper surface; aculeate on the veins below; hispidulous on the veins below; hispidulous
on the intervenium below; with nonbreakaway hairs. Central leaf-blade lobe entire or pinnatifid;
elliptic to broadly elliptic in outline; broadly acute to obtuse at the apex; 3-4 X 1-3 cm. Lateral
leaf-blade lobes asymmetrical; entire; broadly elliptic to transversely elliptic in outline; broadly
acute to obtuse at the apex; 0.5-2 X 0.5-1.5 cm. *Tendrils*: Present; solitary; simple; 1.5-2 cm
long; not aculeate; basally hispidulous. *Inflorescences*: Unisexual. *Male inflorescences*:
Fasciculate or a solitary flower; 1-3-flowered; sessile. *Male flowers*: Pedicel terete in cross
section; 5-11 mm long; hispidulous; without bracteoles. Hypanthium campanulate; 4.8-5 mm
long; 3.2-3.5 mm in diam; hispidulous. Calyx lobes narrowly oblong or narrowly triangular in
outline; narrowly acute at the apex; 1.4-3 X 0.3-0.5 mm; hispidulous. Corolla infundibular;
sparsely hispidulous or puberulent outside; glabrous inside. Corolla tube 0.8-1.1 mm long;
4-4.8 mm in diam; sparsely hispidulous or puberulent outside. Corolla lobes narrowly obovate
to obovate in outline; mucronate and obtuse at the apex; 4.8-6.4 X 3-4 mm; sparsely hispidulous
or puberulent outside; glabrous inside. Stamens separating from the hypanthium ca 1.6 mm from
the base of the hypanthium. Filaments radially compressed in cross section; ca 0.8 mm long;
0.3-0.4 mm wide; glabrous. Anther thecae 2-2.4 mm long; puberulous. Anther connective
obovate; unilobate; obtuse at apex; papillate at the apex; 0.8-1.2 X 0.6-0.8 mm. Disc
cylindrical; 0.8-1 mm long; 0.7-0.8 mm in diam. *Female inflorescences*: A solitary flower.
Female flowers: Pedicel terete in outline; 6-13 mm long; hispidulous; with nonbreakaway hairs;
cylindrical. The lower 3/4 of hypanthium fused to the ovary. Fused portion of hypanthium
ellipsoid; ca 9.6 mm long; ca 4 mm in diam; aculeate; glabrous. Aculei soft; 1.7-2.1 mm long.
Hyaline bristle at apex of aculeus 1.3-1.9 mm long; 4.2-9.5 times as long as the opaque base.
The upper 1/4 of hypanthium free from the ovary. Free portion of hypanthium 3.5-4 mm long;
3.2-4 mm in diam; not aculeate outside; hispidulous outside; minutely puberulent inside. Calyx
lobes linear to narrowly oblong in outline; narrowly acute at the apex; 2.4-3.5 X 0.3-0.6 mm;
sparsely hispidulous. Corolla sparsely hispidulous outside; minutely puberulent inside. Corolla
tube present; 0.8-1.3 mm long; 4-4.8 mm in diam; sparsely hispidulous outside. Corolla lobes

broadly elliptic or ovate in outline; broadly acute and mucronate at the apex; 4.6-8.5 X 3.8-5.4 mm; sparsely hispidulous outside; minutely puberulent inside. Staminodes present; separating from the free portion of the hypanthium ca 1.2 mm above the ovary; ca 0.5 mm long; ca 0.2 mm in diam; glabrous. Style 1.2-1.4 mm long; ca 0.6 mm in diam; subtended by a circular disc. Disc 0.2-0.6 mm long; 1.3-1.6 mm in diam. Stigma ca 2.4 mm long; ca 2.4 mm in diam; lobate; with 5 or 6 finger-like projections on the margin. Finger-like stigmatic projections ca 1.6 mm long; ca 0.8 mm in diam; papillate. *Fruit*: Not geocarpic, maturing above ground and readily visible. Pedicel sulcate in cross section; 1-3 cm long; hispidulous; with nonbreakaway hairs; cylindrical. Fruit bicolored with longitudinal stripes from base to apex (when immature) and monocolored (when mature); dark green (when immature) or yellow (when mature); with light green longitudinal stripes (when immature); globose to ellipsoid; 2.5-4 cm long; 2-3.5 cm in diam; sparsely aculeate; glabrous; blunt at the apex. Aculei terete; with hyaline bristle lost; 0.5-1.6 mm long; 0.3-1.8 mm in diam. *Seeds*: Elliptic; 5-6.4 X 2.6-3 mm; 1.4-1.6 mm thick; unwinged. *Chromosome number*: $2n = 24$ (Deakin, Bohn & Whitaker, 1971; Ramachandran, 1984; Yadava, Singh & Arya, 1984; Ramachandran & Narayan, 1985) or 48 (Dane & Tsuchiya, 1979; Yadava, Singh & Arya, 1984; Nijs & Visser, 1985).

   Distribution: Africa: Ethiopia, Kenya, Rwanda, Tanzania, Uganda, and Zaire (easternmost areas near the Rift Valley).

   This species is endemic to tropical east Africa. Naudin (1859) cited it as a synonym of *C. figarei* Delile, but subsequent authors (Cogniaux, 1881; Cogniaux and Harms, 1924; Jeffrey, 1965, 1967; Kerauden-Aymonin, 1975; Halliday and Jeffrey, 1978) have accepted it as a good species. Meeuse (1962) also accepted this species, but he broadened its concept to include *C. meeusei* and *C. pustulatus* (Jeffrey, 1965).

   Meiotic analysis and pollen fertility of F_1 hybrids (Rammsdonk et al., 1989) suggest a strong link with *C. anguria*, but chloroplast DNA studies (Perl-Treves and Galun, 1985) have shown its greatest similarity to be with *C. pustulatus*, from which it differs in aculeate stems and petioles. It also exhibits strong morphological similarity to *C. aculeatus*, from which it is easily distinguished by the characteristics summarized in couplet 35 of the key to species of *Cucumis* and by its isozyme composition (Staub et al.,1991).

18. **Cucumis aculeatus** Cogniaux, Bot. Jahrb. Syst. 21: 209. 1896.—TYPE: TANZANIA.
       Moshi District: Kilimanjaro, 1200 m, 8 III 1894, *G.L.A. Volkens 1972* (lectotype: BR).

   Plants: Herbs; perennial; with a woody rootstock; lacking tubers; monoecious. Stems procumbent or climbing; sulcate (strongly so when young and weakly so when older); aculeate (with the retrorse aculei 1.5-2.0 mm long, distinctly different from hairs); glabrous or hispid (rarely); with nonbreakaway hairs. Stem hairs 0.3-0.9 mm long. Nodes not geniculate. Internodes 3-7 cm long. *Leaves*: Petioles 1-3 cm long; sulcate; aculeate; glabrous. Leaf blades 3- or 5-palmately (rarely) lobed or trilobate (rarely); with the margin serrate; ovate to broadly ovate in outline; cordate to subtruncate at the base; with or without a basal sinus; 0-0.8 cm deep; acute to broadly acute at the apex; 5-10 X 3.5-7 cm; 0.8-1.5 times longer than wide; hispid on the upper surface; not aculeate on the veins below; hispid on the veins below (with scattered aculei on the main vein); hispid on the intervenium below; with nonbreakaway hairs. Central leaf-blade lobe entire; ovate to broadly ovate in outline; acute to broadly acute at the apex; 3-4(-7) X 2-4 cm. Lateral leaf-blade lobes asymmetrical; entire; shallowly ovate to ovate in outline; broadly acute at the apex; 1-2 X 1.5-2 cm. *Tendrils*: Present; solitary; simple; 2-10 cm long; aculeate (on basal 1/3-1/2); glabrous. *Inflorescences*: Unisexual. *Male inflorescences*: A solitary flower, or racemose (rarely); 1- or 2-flowered; sessile. *Male flowers*: Pedicel terete in cross section; 3-6.5 mm long; hispid (to densely so); without bracteoles. Hypanthium narrowly campanulate; 3.2-3.8(-4.8) mm long; 2.5-2.8 mm in diam; hispidulous (to weakly so). Calyx lobes narrowly oblong in outline; narrowly acute at the apex; 1.5-2.4 X ca 0.3 mm; hispidulous to glabrate. Corolla infundibular; glabrous to hispidulous (sparsely) outside; glabrous inside. Corolla tube ca 0.5 mm long; ca 2.5 mm in diam; glabrous outside. Corolla lobes obovate to narrowly obovate in outline; acute to broadly acute at the apex; 5.2-6.5 X 1.9-4.8 mm; glabrate or glabrous (with the apical 1/3 sparsely hispidulous to glabrate) outside; glabrous inside. Stamens separating from the hypanthium ca 1.5 mm from the base of the hypanthium. Filaments terete in cross section; ca 0.4 mm long; ca 0.2 mm wide; glabrous. Anther thecae 2-2.5 mm long; glabrous. Anther connective oblong to narrowly oblong; unilobate or bilobate (at the apex or apices); obtuse at apex; smooth at the apex; 0.4-1.4 X ca 0.2 mm. Disc cylindrical; ca 0.6 mm long; ca 0.8 mm in diam. *Female inflorescences*: A solitary flower. *Female flowers*: Pedicel terete in outline; 4-12 mm long; hispid (to sparsely so and rarely with short bristle-tipped aculei among the hairs); with nonbreakaway hairs; cylindrical. The lower 2/3 of hypanthium fused to the ovary. Fused portion of hypanthium ovoid; ca 9.5 mm long; ca 4 mm in diam; densely aculeate; glabrous. Aculei stiff or soft (rarely); 1.4-2 mm long. Hyaline bristle at apex of aculeus 0.6-1.4 mm long; 0.8-2.3 times as long as the opaque base. The upper 1/3 of hypanthium free from the ovary. Free portion of hypanthium 3.2-4 mm long; ca 4 mm in diam; not aculeate outside; hispidulous outside; glabrous inside. Calyx lobes narrowly oblong in outline; narrowly acute at the apex; 1.2-2 X ca 0.4 mm; hispidulous to glabrate. Corolla hispidulous (on the veins) outside; glabrous inside. Corolla tube present; 0.4-0.8 mm long; ca 0.5 mm in diam; glabrous outside. Corolla lobes broadly obovate to broadly elliptic in outline; obtuse at the apex; 4.5-5.6 X 4-4.8 mm; hispidulous (on the veins) outside; glabrous inside. Staminodes present; separating from the free portion of the hypanthium ca 1.2 mm above the ovary; 0.5-1.2 mm long; ca 0.3 mm in diam; glabrous. Style ca 1.2 mm long; ca 0.8 mm in diam; subtended by a circular disc. Disc ca 0.6 mm long; ca 1.8 mm in diam. Stigma ca 2.5 mm long; ca 3 mm in diam; lobate; with 6 finger-like projections on the margin. Finger-like stigmatic projections 0.5-1.2 mm long; 0.5-0.6 mm in diam. *Fruit*: Not geocarpic, maturing above ground and readily visible. Pedicel sulcate in cross section; 1.5-2.5 cm long; glabrate; with nonbreakaway hairs; cylindrical. Fruit monocolored; yellow; ellipsoid (to narrowly so); 6-8 cm long; 4-5 cm in diam; aculeate (but many aculei lost from mature fruit); glabrous; blunt at the apex. Aculei terete; with hyaline bristle lost; 0.3-1.2 mm long; 0.4-1.2 mm in diam. *Seeds*: Elliptic; 5.6-6.8 X 2.7-3.2 mm; 1.3-1.6 mm thick; unwinged. *Chromosome number*: 2n = 48 (Dane et al., 1980).

Distribution: Africa: rarely from southern Ethiopia, Kenya, Rwanda, Tanzania, Uganda, and eastern Zaire.

Meiotic analysis and pollen fertility of F_1 hybrids (Rammsdonk et al., 1989) demonstrate links with *C. anguria*, *C. zeyheri*, *C. ficifolius*, and *C. pustulatus*. The species shows greatest morphological similarity to *C. ficifolius* from which it is separated by the characters given in couplet 35 of the Key to the Species of *Cucumis* and by its isozyme composition (Staub et al., 1991). Dane et al. (1980) obtained hybrids between *C. aculeatus* and *C. ficifolius*. One set of *C. aculeatus* chromosomes paired with those of *C. ficifolius* indicating that *C. ficifolius* is one of its probable parents. From the other three species, *C. aculeatus* is easily distinguished by its aculeate stems and petioles.

19. **Cucumis pustulatus** Naudin ex Hooker filius in Oliver, Fl. trop. Afr. 2: 544. 1871.—TYPES: ETHIOPIA. 31 Jun 1862, *G.H.W. Schimper 835* (lectotype [Jeffrey, 1967]: K!; isolectotype: BM!); *Roth* (syntype: not seen); Kouka, *Vogel* (syntype: not seen).

62   Taxonomy

Cucumis figarei Delile ex Naudin var. *ficifolius* Naudin, Ann. Sci. Nat. Bot., sér. 4, 11:
17. 1859.—TYPES: ETHIOPIA. Daroe, 1 Sep 1853, *G.H.W. Schimper 1423*
(lectotype: P! [Kew negative 4336]; isolectotype: P!); Choaché, Melota, 13 Aug
1852, *G.H.W. Schimper 412* (syntypes: LE! P!).—SAUDI ARABIA. Arabia
Felici, *P.E. Botta* (syntype: not seen).

Plants: Herbs; perennial; with a woody rootstock; lacking tubers; monoecious. Stems
procumbent; sulcate; not aculeate; hispidulous or retrorse-strigose; with nonbreakaway hairs.
Stem hairs 0.3-1.3 mm long. Nodes not geniculate. Internodes 2.5-6.5 cm long. *Leaves*: Petioles
2-4(-7.5) cm long; sulcate; not aculeate; pubescence a single type on each petiole; hispidulous or
retrorse-strigose; with nonbreakaway hairs. Leaf blades 3- or 5-palmately lobed or entire
(rarely); with the margin serrate; broadly ovate in outline; cordate at the base; with a basal
sinus; 0.5-1(-1.5) cm deep; broadly acute or obtuse at the apex; 2-5.5(-9) X 2-5(-8) cm; 1.1-1.2
times longer than wide; hispidulous on the upper surface; not aculeate on the veins below;
hispidulous on the veins below; hispidulous on the intervenium below; with nonbreakaway
hairs. Central leaf-blade lobe entire or pinnatifid; broadly ovate to shallowly ovate, or broadly
elliptic to elliptic in outline; broadly acute or obtuse at the apex; 1.5-3(-4.5) X 1.2-3(-5) cm.
Lateral leaf-blade lobes asymmetrical; entire or pinnatifid; shallowly ovate to very shallowly
ovate, or broadly elliptic in outline; broadly acute or obtuse at the apex; 0.4-1(-2) X
0.7-2.5(-3.5) cm. *Tendrils*: Present; solitary; simple; 2-4(-5.5) cm long; not aculeate;
hispidulous or antrorse-strigose. *Inflorescences*: Unisexual. *Male inflorescences*: A solitary
flower or fasciculate (rarely); 1- or 2-flowered; sessile. *Male flowers*: Pedicel terete in cross
section; 6-24 mm long; hispidulous or antrorse-strigose; without bracteoles. Hyanthium
campanulate; 3.2-4.8(-5.2) mm long; 3.2-4 mm in diam; hispidulous (to sparsely so). Calyx
lobes narrowly oblong or narrowly triangular in outline; narrowly acute at the apex; 1-2(-2.8) X
0.3-0.8 mm; hispidulous (to sparsely so). Corolla infundibular; shortly hispidulous outside;
puberulent inside. Corolla tube 0.6-1 mm long; ca 4.4 mm in diam; shortly hispidulous outside.
Corolla lobes elliptic to broadly elliptic in outline; obtuse and mucronate at the apex;
4.8-6.8(-10) X 3.6(-4.8) mm; shortly hispidulous outside; puberulent inside. Stamens separating
from the hypanthium 1.4-1.6 mm from the base of the hypanthium. Filaments terete in cross
section; 0.4-0.6 mm long; ca 0.3 mm wide; glabrous. Anther thecae 2.4-2.6 mm long;
puberulent. Anther connective obovate; unilobate; obtuse at apex; papillate at the apex; ca 1.2 X
0.6-1 mm. Disc cylindrical; 0.6-0.8 mm long; 1-1.7 mm in diam. *Female inflorescences*: A
solitary flower. *Female flowers*: Pedicel terete or sulcate in outline; 15-35(-45) mm long;
hispidulous or antrorse-strigose; with nonbreakaway hairs; weakly flaring upwards from a
narrower base to a wider apex. The lower 3/4 of hypanthium fused to the ovary. Fused portion
of hypanthium ellipsoid; 11-12 mm long; 3.2-4.8 mm in diam; aculeate; glabrous. Aculei stiff;
0.7-1.6 mm long. Hyaline bristle at apex of aculeus 0.3-0.4 or 0.8-1 mm long; 0.7-1 or 2.2-2.5
times as long as the opaque base. The upper 1/4 of hypanthium free from the ovary. Free
portion of hypanthium ca 4 mm long; ca 5.6 mm in diam; not aculeate outside; sparsely
hispidulous outside; sparsely puberulent inside. Calyx lobes narrowly oblong in outline;
narrowly acute at the apex; 0.8-1 or 3.2-4.8 X 0.2-0.8 mm; puberulent. Corolla sparsely
hispidulous outside; sparsely puberulent inside. Corolla tube present; ca 0.8 mm long; ca 4.8
mm in diam; sparsely hispidulous outside. Corolla lobes broadly elliptic in outline; obtuse and
mucronate at the apex; 5.6-6 or 11 X 5.2 or 8 mm; sparsely hispidulous outside; sparsely
puberulent inside. Staminodes present; separating from the free portion of the hypanthium ca 1.2
mm above the ovary; ca 0.7 mm long; ca 0.3 mm in diam; glabrous. Style ca 1.2 mm long; ca 1
mm in diam; subtended by a circular disc. Disc ca 0.6 mm long; ca 2.3 mm in diam. Stigma ca
3.2 mm long; ca 2.8 mm in diam; lobate; with 6 finger-like projections on the margin.

Finger-like stigmatic projections 1.8-2.2 mm long; 0.6-0.8 mm in diam; papillate. *Fruit*: Not geocarpic, maturing above ground and readily visible. Pedicel sulcate in cross section; 2.5-10 cm long; hispidulous; with nonbreakaway hairs; flaring upwards from a narrower base to a wider apex. Fruit monocolored (when mature) or bicolored with longitudinal stripes from base to apex (when immature); greenish yellow (when mature); with pale light green longitudinal stripes; ellipsoid to cylindrical; 5.5-8.5 cm long; 3.5-5 cm in diam; sparsely aculeate; glabrous; blunt at the apex. Aculei terete; with hyaline bristle lost; 0.9-1.6 or 3.2-6.4 mm long; 1.4-2.4 or 4.8-5.6 mm in diam. *Seeds*: Elliptic; 5.2-6.4 X 2.7-3 mm; 1.1-1.4 mm thick; unwinged. *Chromosome number*: 2n = 24 (Shimotsuma, 1965), 48 (Ramachandran, 1984; Nijs & Visser, 1985; Ramachandran & Narayan, 1985), or 72 (Dane & Tsuchiya, 1979).

Distribution: Africa: Chad, Ethiopia, Kenya, Niger, Nigeria, Sudan, Tanzania, and Uganda. Southwest Asia: Saudi Arabia and Yemen.

When Hooker (1871) published *C. pustulatus*, he cited *C. abyssinicus* A. Richard as a synonym of both *C. figarei* and *C. pustulatus*. The type of *C. abyssinicus* is *Quartin-Dillon sn*, Chiré, Ethiopia (Richard, 1847), and since Hooker (1871) cited it under *C. figarei*, his *C. pustulatus* is legitimate because he did not include the type of *C. abyssinicus* within its circumscription. Therefore, the lectotype chosen by Jeffrey (1967) stands.

Traditionally this species has been referred to as "*Cucumis figarei* Naudin." Delile first published the epithet *figarei* in 1843 as "*Cucumis figarei* nov. sp.," a nomen nudum without any text, and he published it again in 1847 as "*Cucumis figarei* Delile," again as a nomen nudum (L. Bertrand, pers. comm.). Naudin (1859) accepted the species in his monograph, and said, "Il y a plus de vingt ans que le *C. figarei* est cultivé à Montpellier, où il paraît avoir été introduit par le professeur Delile. Nous avons dù lui conserver le nom que ce botaniste lui avait donné, et lequel il l'annonçait tous les ans dans les catalogues du Jardin botanique de cette ville." He also cited in synonymy under *C. figarei*, *C. ficifolius* A. Richard and *C. abyssinicus* A. Richard; therefore the name is illegitimate because the types of both species were included.

Following chloroplast DNA analysis (Perl-Treves and Galun, 1985) that showed different collections of *C. pustulatus* to be similar in their structures to *C. ficifolius* and *C. meeusei*, I have placed it in a position intermediate between these two species. *Cucumis pustulatus* is easily separated from *C. ficifolius* by lack of aculei on the stems and petioles; the differences between it and *C. meeusei* are set forth in couplet 34 of the key to species of *Cucumis*.

20. **Cucumis meeusei** C. Jeffrey, Kew Bull. 19(2): 218. 1965.—TYPES: NAMIBIA. Bed of Omuramba Omatako at Karahuwisa, 27 Nov 1955, *B. de Winter 3713* (holotype: K; isotypes: M! PRE!); between Hoffnungsfelde and Haris, 8 Jan 1916, *H.H.W. Pearson 9526* (paratype: BOL!); Kaokoveld, Herero, Oukungu, below houses at Ohopoho, 24 Mar 1954, *B. de Winter & O.A. Leistner 5157* (paratypes: K! M!). South Africa, Cape Province: 3½ miles north of Houtkraal Stn. Central, upper Karroo, alt. ca 1280 m, 17 Apr 1956, *J.P.H. Acocks 18831* (paratypes: K! L! MO! PRE!); Kimberley, 19 miles WSW of Kimberley on Kloedoesberg Rd., alt. ca 1220 m, *O.A. Leistner 1230* (paratype: M!).—CULTIVATED. United States, California, La Jolla, U.S. Horticultural Field Station, 15 Aug 1962, *T.W. Whitaker s.n.* (paratype: K!).

Plants: Herbs; perennial; with a woody rootstock; lacking tubers; monoecious. Stems procumbent; sulcate; not aculeate; antrorse-strigose or retrorse-strigose (to weakly so); with nonbreakaway hairs. Stem hairs 0.4-0.8 mm long. Nodes not geniculate. Internodes 3-6(-8) cm long. *Leaves*: Petioles 1-2.5 cm long; sulcate; not aculeate; pubescence a single type on each petiole; antrorse-strigose or retrorse-strigose (to weakly so); with nonbreakaway hairs. Leaf blades 5-palmately lobed; with the margin weakly serrate; ovate to broadly ovate in outline; subcordate at the base; with a basal sinus; 0.2-0.6 cm deep; acute to broadly acute at the apex; 3.5-8.5(-11) X 3-7 cm; 1.2-1.6 times longer than wide; antrorse-strigose or scabrous on the upper surface; not aculeate on the veins below; antrorse-strigose (to weakly so) on the veins below; scabrous or antrorse-strigose on the intervenium below; with nonbreakaway hairs. Central leaf-blade lobe entire or pinnatifid; narrowly elliptic to elliptic in outline; acute to broadly acute at the apex; 2.5-6.5(-7.5) X 0.2-2.5(-3) cm. Lateral leaf-blade lobes asymmetrical; entire; elliptic in outline; acute to broadly acute at the apex; 0.5-2.5 X 0.4-1.8 cm. *Tendrils*: Present; solitary; simple; 0.5-3(-5) cm long; not aculeate; antrorse-strigose (to weakly so) or retrorse-strigose. *Inflorescences*: Unisexual. *Male inflorescences*: A solitary flower; sessile. *Male flowers*: Pedicel terete in cross section; 4-6 mm long; antrorse-strigose, or retrorse-strigose; without bracteoles. Hypanthium infundibular or campanulate; ca 4.4 mm long; 2.8-3.6 mm in diam; hispid. Calyx lobes triangular in outline; narrowly acute at the apex; 1.6-2.4 X 0.8-1 mm; hispidulous. Corolla infundibular; puberulent outside; puberulent inside. Corolla tube 0.8-1.2 mm long; 3.6-4.4 mm in diam; puberulent outside. Corolla lobes broadly elliptic in outline; mucronate and obtuse at the apex; 4-5.2 X 3.2-4 mm; puberulent outside; puberulent inside. Stamens separating from the hypanthium ca 1.6 mm from the base of the hypanthium. Filaments terete in cross section; 0.4-0.8 mm long; 0.3-0.4 mm wide; glabrous. Anther thecae 2.2-2.8 mm long; puberulent. Anther connective obovate; unilobate; obtuse at apex; papillate at the apex; 1.2-1.6 X 0.6-1 mm. Disc cylindrical; 0.6-0.8 mm long; 1.4-1.6 mm in diam. *Female inflorescences*: A solitary flower. *Female flowers*: Pedicel terete in outline; 8-13 mm long; antrorse-strigose or retrorse-strigose; with nonbreakaway hairs; weakly flaring upwards from a narrower base to a wider apex. The lower 2/3 of hypanthium fused to the ovary. Fused portion of hypanthium ellipsoid; 7.2-8 mm long; 4-4.8 mm in diam; aculeate; glabrous. Aculei stiff; 1-1.8 mm long. Hyaline bristle at apex of aculeus ca 0.5 mm long; 0.4-0.5 times as long as the opaque base. The upper 1/3 of hypanthium free from the ovary. Free portion of hypanthium 3.2-4 mm long; ca 4 mm in diam; not aculeate outside; short antrorse-strigose outside; puberulent inside. Calyx lobes narrowly oblong in outline; narrowly acute at the apex; 2.4-2.8 X 0.3-0.4 mm; short antrorse-strigose. Corolla puberulent outside; puberulent inside. Corolla tube present; 0.4-0.8 mm long; ca 4 mm in diam; puberulent outside. Corolla lobes elliptic in outline; mucronate and obtuse at the apex; 6.4-8 X 4-4.8 mm; puberulent outside; puberulent inside. Staminodes present; separating from the free portion of the hypanthium 1-1.2 mm above the ovary; 0.6-0.8 mm long; ca 0.2 mm in diam; glabrous. Style 1.2-1.6 mm long; ca 0.8 mm in diam; subtended by a circular disc. Disc ca 0.4 mm long; 2.2-2.3 mm in diam. Stigma 2.4-4 mm long; ca 3.2 mm in diam; lobate; with 5 or 6 finger-like projections on the margin. Finger-like stigmatic projections 1.2-2.4 mm long; ca 1.2 mm in diam; papillate. *Fruit*: Not geocarpic, maturing above ground and readily visible. Pedicel terete in cross section; 1.5-4(-6) cm long; antrorse-strigose or retrorse-strigose; with nonbreakaway hairs; weakly flaring upwards from a narrower base to a wider apex. Fruit monocolored; green (when young) or yellow (when mature); ellipsoid to globose; 4.5-6 cm long; 3.5-4 cm in diam; aculeate (the aculei conical); glabrous; blunt at the apex. Aculei terete; with hyaline bristle lost; 0.5-3.5 mm long; 1.5-4 mm in diam. *Seeds*: Elliptic; 5.6-6.7 X 2.7-3.2 mm; 1.4-1.6 mm thick; unwinged. *Chromosome number*: 2n = 48 (Ramachandran, 1984; Yadava, Singh & Arya, 1984; Ramachandran & Narayan, 1985).

Distribution: Africa: northern Botswana, northern Namibia, and South Africa (northern Cape Province).

Cucumis meeusei is endemic to the northern half of the Kalahari-highveld regional transition zone as defined by White (1983). Many of the collections are outside of the area of Kalahari sands.

The pubescence on stems, petioles, leaf blades, tendrils, and both flower and fruit pedicels, is either antrorse-strigose or retrorse-strigose with hairs tightly appressed in some specimens and mixed, loosely appressed and suberect hairs in others. Aculei on the fruit of those with antrorse-strigose pubescence tend to be more variable in size and among the largest found on the species.

Chloroplast DNA analysis (Perl-Treves and Galun, 1985) shows *C. meeusei* and *C. pustulatus* to be very close, but they can be easily separated by the characters given in couplet 34 of the Key to the Species of *Cucumis*.

21. **Cucumis jeffreyanus** Thulin, Nord. J. Bot. 11(5): 539. 1991.—TYPES: ETHIOPIA. HARERGE REGION: N of Erer-Gota near Idorra, 9°56'N, 41°20'E, 31 Aug 1963, *W. Burger 3214* (holotype: K).—SOMALIA. BAY REGION: 7 km W of Baydhabo on road to Luuq, 3°10'N, 45°36'E, 19 May 1983, *J.B. Gillett & C.F. Hemming 24201* (paratypes: EA, K); Diinsoor, 2°25'N, 42°59'E, 21 May 1990, *M. Thulin, Hedrén & Abdi Dahir 7648* (paratype: UPS).

Plants: Herbs; probably perennial; monoecious. Stems procumbent or climbing (green); sulcate; not aculeate; retrorse-strigose (scabrous, fide Thulin, pers. comm.); with nonbreakaway hairs. Stem hairs 0.2-0.8 mm long. Nodes not geniculate. Internodes 2.5-9 cm long. *Leaves*: Petioles 1-4 cm long; sulcate; not aculeate; pubescence a single type on each petiole; retrorse-strigose (scabrous, fide Thulin, pers. comm.); with nonbreakaway hairs. Leaf blades 3- or 5-palmately lobed; with the margin serrate; ovate to broadly ovate in outline; cordate at the base; with a basal sinus; 0.5-1 cm deep; acute to broadly acute, or obtuse at the apex; 1.5-5.5 X 1.8-4.6 cm; 1.2-1.3 times longer than wide; sparsely short antrorse-strigose on the upper surface (scabrous, fide Thulin, pers. comm.); not aculeate on the veins below; antrorse-strigose on the veins below (scabrous, fide Thulin, pers. comm.); short antrorse-strigose on the intervenium below (denser than above, scabrous, fide Thulin, pers. comm.); with nonbreakaway hairs. Central leaf-blade lobe entire; elliptic in outline; acute to broadly acute, or obtuse at the apex; 1.5-4 X 1-2 cm. Lateral leaf-blade lobes asymmetrical; entire; broadly elliptic or very broadly ovate to broadly ovate in outline; broadly acute to obtuse at the apex; 0.5-1.8 X 0.5-1.5 cm. *Tendrils*: Present; solitary; simple; 1.5-4 cm long; not aculeate; retrorse-strigose (on the basal half). *Inflorescences*: Unisexual. *Male inflorescences*: A solitary flower or fasciculate (rarely); 1- or 2-flowered; sessile. *Male flowers*: Pedicel terete in cross section; 5-20 mm long; hispid, or retrorse-strigose to hispidulous; without bracteoles. Hypanthium campanulate; 3-6 mm long; ca 2.4 mm in diam; hispidulous to hispid. Calyx lobes linear or narrowly triangular in outline; narrowly acute at the apex; 1.3-2 X 0.4-0.6 mm; sparsely hispidulous to hispid. Corolla hirsute (on the veins) outside; puberulent inside. Corolla lobes elliptic to oblong in outline; ca 10 mm long; densely puberulent on the intervenium outside and hirsute or antrorse-strigose on the veins outside; puberulent inside. Stamens separating from the hypanthium 1.2-2.8 mm from the base of the hypanthium. Filaments radially compressed in cross section; 0.4-0.6 mm long; ca 0.3 mm wide; glabrous. Anther thecae ca 2.6 mm long; puberulent. Anther connective obovate; unilobate; obtuse at apex; papillate at the apex; 0.9-1 X 0.3-0.5 mm. Disc cylindrical; ca 0.5 mm long; ca 1.1 mm in diam. *Female inflorescences*: A solitary flower. *Female flowers*: Pedicel terete in outline; 5-12 mm long; short retrorse-strigose, scabrous, or hirsute; with nonbreakaway hairs; cylindrical. The lower 2/3 of hypanthium fused to the ovary. Fused portion of hypanthium ellipsoid; 6-7 mm long; 2.8-5 mm in diam; aculeate; glabrous. Aculei soft; 1.6-2.2 mm long. Hyaline bristle at apex of aculeus 1-1.2 mm long; 1-2.1 times as long as the opaque base. The upper 1/3 of hypanthium free from the ovary. Free portion of hypanthium 3.2-4 mm long; ca 2.8 mm in diam; not aculeate outside; sparsely hispidulous to hispid outside; puberulent inside. Calyx lobes linear to narrowly oblong in outline; narrowly acute at the apex; 1.4-2.5 X 0.2-0.4 mm; short antrorse-strigose or hispid. Corolla hirsute outside. Corolla tube present; hirsute outside. Corolla lobes elliptic in outline; obtuse and mucronate at the apex; 4-5.5 mm; short antrorse-strigose or hirsute outside. Staminodes present (sometimes very rudimentary); separating from the free portion of the hypanthium ca 1.2 mm above the ovary; ca 0.4 mm long; glabrous. Style 0.5-0.9 mm long; ca 0.4 mm in diam; subtended by a circular disc. Disc ca 0.5 mm long; ca 1.6 mm in diam. Stigma ca 2.4 mm long; ca 2.5 mm in diam; lobate; with 6 finger-like projections on the margin. Finger-like stigmatic projections ca 1.6 mm long; ca 0.8 mm in diam; papillate. *Fruit*: Not geocarpic, maturing above ground and readily visible. Pedicel sulcate in cross section; 0.5-3.5 cm long; retrorse-strigose (when young) or glabrate (when older); with nonbreakaway hairs; cylindrical. Fruit bicolored with longitudinal stripes from base to apex; pale green (with yellow or orange aculei) or white; with light green longitudinal stripes; ellipsoid to globose; 3-4 cm long; 2.5-3.5 cm in diam; aculeate; glabrous; blunt at the apex. Aculei terete; with hyaline bristle lost; 3-8.8 mm long; 1.6-2.4 mm in diam. *Seeds*: Elliptic; 3.2-4.7 X 1.6-2.2 mm; 1-1.1 mm thick; unwinged.

    Distribution: Africa: Ethiopia, Kenya, and Somalia.

    Cucumis jeffreyanus was mentioned by Jeffrey as an undescribed species in the *Flora of Tropical East Africa* (1967) and as species *A* in the manuscript for the *Flora of Ethiopia* (pers. comm.). In addition to the collections designated as types by Thulin, cited above, the following collections pertain to this species: *Ash 2505* (K! MO!), *Kokwaro, Mattrenge & Mburi 3553* (K!), and *Makin EA12994* (K!).

22. **Cucumis hastatus** Thulin, Nord. J. Bot. 11(5): 535. 1991.—TYPES: SOMALIA. BAY REGION: 52 km from Yaaq Bari Weyne to Buurhakaba, 2°45'N, 44°15'E, 23 May 1990, *M. Thulin, Hedrén & Abdi Dahir 7765* (holotype: UPS; isotypes: K MOG); 48 km W of Baydhabo on road to Luuq, 3°15'N, 43°10'E, 19-20 May 1983, *J.B. Gillett & C.F. Hemming 24229* (paratype: K); Baydhabo, Bonka, 3°9'N, 43°40'E, 5 Jun 1987, *Eagleton 19* (paratype: K); 3°8'N, 43°42'E, 19 Mar 1982, *J.J. Beckett & White 1810* (paratype: EA); 3 km S of Baydhabo on road to Diinsoor, 3°5'N, 43°38'E, 11 Jun 1983, *J.B. Gillett & C.F. Hemming 24650* (paratype: EA, K); 68 km from Baydhabo to Baardheere, 2°58'N, 43°6'E, 26 May 1987, *I. Friis, K. Vollesen & Hassan 4794* (paratype: K); 80 km on road from Baydhabo to Diinsoor, 2°40'N, 43°14'E, 19 May 1990, *M. Thulin, Hedrén & Abdi Dahir 7579* (paratype: UPS); 24 km on road from Diinsoor to Ufurow, 2°37'N, 42°56'E, 22 May 1990, *M. Thulin, Hedrén & Abdi Dahir 7707* (paratypes: E, FT, K, MOG, UPS). BAKOOL REGION: Tayeeglow Distr., 4°14'N, 44°14'E, 1 Aug 1983, *Abukar Sheikh A2041* (paratype: K). GEDO REGION: 4 km W of Luuq, 3°49'N, 42°31'E, 27 May 1988, *Somalia Medicinal Plant Project SMP219* (paratype: K, UPS); 10 km SE of Luuq, 10 May 1983, *Wieland 1089* (paratype: UPS); Allengo, 27 Sep 1913, *G. Paoli 859* (paratype: FT); escarpment 17 km from Baardheere near road to Fafadun, 2°22'N, 42°8'E, 4 Jun 1989, *M. Thulin & Bashir Mohamed 6892* (paratype: K, UPS). HIIRAAN REGION: 1.6 km N of Halgen, 24 Nov 1983, *Kuchar & Abdihakeem 15573* (paratype: UPS); 5.5 km S of Dharyo on road to Gobyaxas, 3°37'N, 44°54'E, 12 Aug 1987, *Kuchar 17327* (paratype: K).

    Plants: Herbs; perennial; monoecious. Stems procumbent or climbing (older parts pale brown to whitish, younger parts pale green); sulcate; not aculeate; scabrous (a mixture of more robust more or less retrorsely hooked hairs and slender short more or less straight hairs); with nonbreakaway hairs. Nodes not geniculate. Internodes 4-5 cm long. *Leaves*: Petioles 0.4-1.2 cm long; not aculeate; pubescence a single type on each petiole; scabrous; with nonbreakaway hairs. Leaf blades trilobate; with the margin serrate; ovate to triangular in outline; truncate or subcordate at the base; with or without a basal sinus; acute to obtuse at the apex; 1.2-6 X 1.2-6.5 cm; scabrous on the upper surface; not aculeate on the veins below; scabrous on the veins below; scabrous on the intervenium below (paler green and denser than above); with nonbreakaway hairs. Central leaf-blade lobe entire; ovate in outline; acute to obtuse at the apex; 1.6-4.2 X 0.8-1.8 cm. Lateral leaf-blade lobes asymmetrical; entire; ovate to broadly ovate in outline; acute to obtuse at the apex; 0.6-1.4 X 0.4-1.6 cm. *Tendrils*: Present; solitary; simple; 4-10 cm long; not aculeate; scabrous. *Inflorescences*: Unisexual. *Male inflorescences*: A solitary flower; sessile. *Male flowers*: Pedicel 1.5-4(-10) mm long; scabrous; without bracteoles. Hypanthium campanulate; 3.5-6.5 mm long; scabrous. Calyx lobes linear in outline; weakly acute at the apex; 1.2-3.2 mm long; scabrous. Corolla infundibular; puberulent inside. Corolla lobes elliptic to oblong in outline; weakly acute at the apex; 3.5-7.2 mm long; scabrous (on the veins) outside; puberulent inside. Stamens separating from the hypanthium 1.5-3 mm from the base of the hypanthium. Anther thecae puberulous. Anther connective obovate; unilobate; obtuse at apex; crenulate at the apex; ca 0.8 mm long. Disc cylindrical. *Female inflorescences*: A solitary flower. *Female flowers*: Pedicel 3-12 mm long; hirsute; cylindrical. The lower 2/3 of hypanthium fused to the ovary. Fused portion of hypanthium ellipsoid; 7-8 mm long; 3-4 mm in diam; densely aculeate; glabrous. The upper 1/3 of hypanthium free from the ovary. Free portion of hypanthium ca 4 mm long; not aculeate outside; hirsute outside. Calyx lobes linear in outline; ca 4 mm long; hirsute. Corolla hirsute outside. Corolla tube present; hirsute outside. Corolla lobes elliptic to oblong in outline; acute at the apex; ca 3.5 mm long; hirsute outside. Staminodes present; glabrous. Style subtended by a circular disc. Stigma ca 2.5 mm in diam; sublobate; with 3 to 4 finger-like projections on the margin. *Fruit*: Not geocarpic, maturing above ground and readily visible. Pedicel 0.8-2.5 cm long; cylindrical. Fruit greenish yellow (when mature); broadly ellipsoid; 4-6 cm long; 2.7-5 cm in diam; sparsely aculeate; glabrous; blunt at the apex. Aculei terete; with hyaline bristle lost; 0.5-1.2 mm long. *Seeds*: Elliptic; 5.5-6 X 2.5-3 mm; ca 1.5 mm thick; unwinged.

Distribution: Africa: southern Somalia.

23. **Cucumis rigidus** E. Meyer ex Sonder in Harvey & Sonder, Fl. cap. 2: 497. 1862.—TYPES: SOUTH AFRICA. Cape Province: On the Gariep [Orange] River, *J.F. Drège s.n.* (lectotype: K[Herbarium Hookerianum]!; isolectotypes: BR[fragments]! K[Herbarium Benthamianum]! L! LE! MO! P! S! W!); Namaqualand, *A. Wyley s.n.* (syntypes: K! MEL!).

Plants: Subshrubs; perennial; with a woody rootstock; lacking tubers; dioecious. Stems erect; sulcate (ridges between the sulci white); not aculeate; antrorse-strigose; with nonbreakaway hairs. Stem hairs 0.2-0.8 mm long. Nodes not geniculate. Internodes 2.5-5.5 cm long. *Leaves*: Petioles (2-)2.5-4.5 cm long; sulcate; not aculeate; pubescence a single type on each petiole; antrorse-strigose; with nonbreakaway hairs. Leaf blades 3- or 5-palmately lobed; with the margin entire; ovate to broadly ovate in outline; truncate to subcordate, or acute at the base; with or without a basal sinus; ca 0.5 cm deep (when present); obtuse at the apex; 3-5 X 3-5.5 cm; 1-1.5 times longer than wide; hispidulous on the upper surface (each hair with a flat base of enlarged, opaque cells, often broken off near base as leaf matures); not aculeate on the veins below; hispidulous on the veins below (each hair with a flat base of enlarged, opaque cells, often broken off near base as leaf matures); hispidulous on the intervenium below (each hair with a flat base of enlarged, opaque cells, often broken off near base as leaf matures); with nonbreakaway hairs. Central leaf-blade lobe entire; elliptic to broadly elliptic in outline; obtuse at the apex; 1.5-3 X 1-2.5 cm. Lateral leaf-blade lobes asymmetrical; entire; elliptic or shallowly ovate in outline; obtuse at the apex; 0.5-1.5 X 1.5-2 cm. *Tendrils*: Absent. *Inflorescences*: Unisexual. *Male inflorescences*: Fasciculate or a solitary flower; 1-4-flowered; sessile. *Male flowers*: Pedicel terete in cross section; 3.6-5.5(-14) mm long; antrorse-strigose; without bracteoles. Hypanthium campanulate; ca 4.4 or 7.2 mm long; ca 4 mm in diam; antrorse-strigose. Calyx lobes linear or narrowly triangular in outline; narrowly acute at the apex; 1.6-2.4 X 0.4 or 0.8-1.2 mm; antrorse-strigose. Corolla infundibular; antrorse-strigose outside; puberulent inside. Corolla tube 0.6-0.8 mm long; 4.4-4.8 mm in diam; antrorse-strigose outside. Corolla lobes broadly elliptic or ovate in outline; obtuse and mucronate at the apex; 2.4-4 or 6-6.8 X 2-2.4 or 4.8-6 mm; antrorse-strigose outside; puberulent inside. Stamens separating from the hypanthium ca 1.2 mm or 2.8 mm from the base of the hypanthium. Filaments radially compressed in cross section; ca 0.8 mm long; ca 0.4 mm wide; glabrous. Anther thecae ca 2.8 mm long; puberulent. Anther connective narrowly oblong or oblong; unilobate; obtuse at apex; papillate at the apex; 0.4-0.8 X ca 0.8 or 1.6 mm. Disc cylindrical; ca 0.4 or 1.2 mm long; ca 1.6 or 2.4 mm in diam. *Female inflorescences*: A solitary flower. *Female flowers*: Pedicel terete in outline; 5-12 mm long; antrorse-strigose; with nonbreakaway hairs; cylindrical. The lower 2/3 of hypanthium fused to the ovary. Fused portion of hypanthium ellipsoid; 6.4-9.6 mm long; 4-4.8 mm in diam; aculeate; glabrous. Aculei stiff; ca 1.2 mm long. Hyaline bristle at apex of aculeus ca 0.6 mm long; ca 1 times as long as the opaque base. The upper 1/3 of hypanthium free from the ovary. Free portion of hypanthium 3.2-4 mm long; ca 4.4 mm in diam; aculeate or not aculeate outside; antrorse-strigose or glabrous outside; puberulent inside. Calyx lobes linear or narrowly triangular in outline; narrowly acute at the apex; 1.2-2 X 0.5-1 mm; antrorse-strigose. Corolla antrorse-strigose outside; sparsely puberulent inside. Corolla tube present; ca 0.4 mm long; ca 4.8 mm in diam; antrorse-strigose outside. Corolla lobes elliptic in outline; obtuse and mucronate at the apex; 2.4-4.8 X 1.6-2 or 2.4-2.8 mm; antrorse-strigose outside; sparsely puberulent inside. Staminodes present; separating from the free portion of the hypanthium ca 0.8 mm above the ovary; 0.8-1.2 mm long; ca 0.4 mm in diam; puberulent. Style ca 0.8 mm long; 0.8-1 mm in diam; subtended by a circular disc. Disc ca 0.6 mm long; ca 2.8 mm in diam. Stigma 2.8-3.2 mm long; 2.8-3.6 mm in diam; lobate; with 5 finger-like projections on the margin. Finger-like stigmatic projections 0.8-1.6 mm long; 0.8-1.2 mm in diam; papillate. *Fruit*: Not geocarpic, maturing above ground and readily visible. Pedicel sulcate in cross section; 0.5-1.3 cm long; antrorse-strigose; with nonbreakaway hairs; cylindrical. Fruit monocolored (fide Meeuse, 1962); ellipsoid or globose; 2.5-4.5 cm long; 2-4 cm in diam; aculeate; glabrous; blunt at the apex. Aculei laterally compressed near the base; with hyaline bristle lost; 3.2-4.5 mm long; 0.6-1 mm in diam. *Seeds*: Seeds elliptic; 5.7-6.7 X 2.7-3.8 mm; 1-1.4 mm thick; unwinged.

Distribution: Africa: Namibia, southernmost area along Orange River, and South Africa, northwestern Cape Province along the Orange River.

Johann Franz Drège was in South Africa from March 1826 to March 1834 (Gunn & Codd, 1981). He traveled extensively and collected more than 200,000 botanical specimens. Upon his return to Europe, the first set of his collections was given to Ernst H.F. Meyer, professor of botany at Königsberg, to be worked up. Unfortunately, Meyer published only two short fascicles of his *Commentariorum de plantis Africae Australioris* in 1836 and 1838 (Stafleu & Cowan, 1981), and left many new species unpublished as *nomina inedita* in Drège's collections.

Arnott (1841) presented a conspectus of the Cucurbitaceae clarifying and amplifying the system of Schrader (1838). *Cucumis rigidus* is given as one of the nine species in the genus and attributed to E. Meyer. As no other information is given, it is a *nomen nudum*.

Naudin (1859) organized his monograph of *Cucumis* in four parts: 1) Introduction; 2) treatment of the genus; 3) treatments of the 13 accepted species; and 4) 2 appendices, *Species vix cognitae, Cucumeribus quaedem adjungendae, reliquae dubii generis*; and *Species a Cucumere remotae aut adhucdum removendae*, i.e., the poorly known species and the excluded species, respectively. *Cucumis rigidus* was placed in the first appendix, and Ecklon was given as its author. Ernst Meyer was cited as the author in Arnott's publication. Naudin described it as being an interesting plant, but because of its striking appearance and lack of tendrils, he was not certain that it was a member of the genus *Cucumis*. He had only examined its female flower, and hoped that other botanists would make further observations on it in the living state, presumably to determine its proper generic position.

Recently, Naudin has been cited as the publishing author of *C. rigidus* (Meeuse, 1962; Roessler, 1968; Gibbs Russell, et al., 1987). Since Naudin did not include it among the accepted species and expressed taxonomic doubt by placing it in his first appendix of poorly known species, he did not validate it as a new species.

In *Flora capensis* (Sonder, 1862), Sonder accepted *C. rigidus* as a good species, provided a detailed description, and cited collections examined. Ernst Meyer is credited as the source of the name, therefore the full author citation is "E. Meyer ex Sonder in Harvey & Sonder." Sonder cited two collections, *Drège* from the Gariep [Orange] River and *Wyley* from Namaqualand, so these are syntypes. The lectotype is designated above from among the duplicates of the *Drège* collection.

This species is known only from the southernmost boundary of Namibia along the Orange River and the northwestern tip of Cape Province, South Africa, along the Orange River south of Namibia. Within that limited range, there are apparently two flower morphs: Longer flowers: Male hypanthium 7.2 mm long, male calyx lobes 4, linear, 1.6-2.4 X 0.8-1.2 mm, male corolla lobes 6.0-6.8 X 4.8-6.0 mm, male flower disc 1.2 mm long, 2.4 mm in diam, female corolla lobes 3.2-4.8 X 2.4-2.8 mm; shorter flowers: Male hypanthium 4.4 mm long, male calyx lobes 5, narrowly triangular, 1.6 X 0.4 mm, male corolla lobes 2.4-4.0 X 2.0-2.4 mm, male flower disc 0.4 mm long, 1.6 mm in diam, female corolla lobes 2.4-3.2 X 1.6-2.0 mm. Male flowers of the longer morph (*Mermüller & Geiss 3655*) exhibit a suite of characters correlated with a hypanthium longer and corolla lobes larger than those of the shorter morph (*Pillans 5501*), while the female flowers of the longer morph (*van Berkel 87*) have only the corolla larger (*Pearson 6911*). Additional material is needed to fully understand this variation and to evaluate its taxonomic significance.

Cucumis rigidus is unique in the genus in lacking stipules and being an erect, dioecious subshrub.

24. **Cucumis baladensis** Thulin, Nord. J. Bot. 11(5): 541. 1991.—TYPES: SOMALIA. SHABEELLAHA DHEXE REGION: 16 km N of Muqdisho along road to Balcad (or Balad), 2°13'N, 45°24'E, 15 May 1990, *Thulin, Hedrén & Abdi Dahir 7464* (holotype: UPS; isotypes: FT K MOG); Balcad Nature Reserve, ca 2 km S of Balcad, 2°20'N, 45°23'E, 25 May 1990, *Thulin, Hedérn & Abdi Dahir 7770* (paratype: UPS); 16 km N of Muqdisho along road to Balcad, 2°13'N, 45°24'E, 25 May 1990, *Thulin, Hedrén & Abdi Dahir 7772* (paratypes: K UPS).

Plants: Herbs; without a woody rootstock; dioecious. Stems procumbent (green, rooting at the nodes); sulcate; not aculeate; puberulent and hirsute. Stem hairs up to 1 mm long. Nodes not geniculate. Internodes 5-7 cm long. *Leaves*: Petioles 1-8 cm long; not aculeate; pubescence a single type on each petiole; hispid. Leaf blades 5-palmately lobed; with the margin serrate;

broadly ovate in outline; cordate at the base; with a basal sinus; 0.5-1 cm deep; acute at the apex; 2-6 X 2-5.5 cm; hirsute on the upper surface; not aculeate on the veins below; hirsute on the veins below (longer than intervenium hairs); hirsute on the intervenium below. Central leaf-blade lobe entire; elliptic in outline; acute at the apex; 2-3 X 1-1.5 cm. Lateral leaf-blade lobes asymmetrical; entire; elliptic in outline; acute at the apex; 1-2.5 X 0.7-1.2 cm. *Tendrils*: Present; solitary; simple; ca 1.5 cm long (pubescent). *Inflorescences*: Unisexual. *Male inflorescences*: A solitary flower; sessile. *Male flowers*: Pedicel 10-20 mm long (pubescent); without bracteoles. Hypanthium campanulate; 5-8 mm long; hirsute. Calyx lobes linear in outline; narrowly acute at the apex; ca 2 mm long; hirsute. Corolla infundibular; puberulent outside; puberulent inside. Corolla tube puberulent outside. Corolla lobes elliptic in outline; acute at the apex; 12-23 mm long; puberulent outside; puberulent inside. Stamens separating from the hypanthium 2-3 mm from the base of the hypanthium. Anther thecae papillose. Anther connective obovate; unilobate; obtuse at apex; crenulate at the apex; 1.5-2 mm long. Disc cylindrical. *Female inflorescences*: A solitary flower. *Female flowers*: Pedicel 11-15 mm long (pubescent); cylindrical. The lower 2/3 of hypanthium fused to the ovary. Fused portion of hypanthium ellipsoid; 8-10 mm long; ca 5 mm in diam (pubescent); not aculeate. The upper 1/3 of hypanthium free from the ovary. Free portion of hypanthium ca 4 mm long; not aculeate outside; hirsute outside. Calyx lobes linear in outline (with a broad base); narrowly acute at the apex; ca 2 mm long; hirsute. Corolla puberulent outside; puberulent inside. Corolla tube present; puberulent outside. Corolla lobes elliptic in outline; obtuse at the apex; ca 16 mm long. Staminodes present (rudimentary). Style ca 1 mm long; subtended by a circular disc. Stigma ca 3.2 mm long; ca 4 mm in diam; lobate; with 6 finger-like projections on the margin. *Fruit*: Not geocarpic, maturing above ground and readily visible. Pedicel ca 2 cm long; cylindrical. Fruit bicolored with longitudinal stripes from base to apex; pale green; with darker green longitudinal stripes; weakly globose; ca 3.5 cm long; ca 3 cm in diam; not aculeate; glabrous (with scattered, stiff hairs with enlarged bases); blunt at the apex. *Seeds*: Elliptic; ca 6.5 X 3.5-3.6 mm; unwinged.

Distribution: Africa: Somalia.

Cucumis ser. **Metuliferi** Kirkbride, ser. nov.—TYPE: *Cucumis metuliferus* E. Meyer ex Naudin. Species 25 and 26.

A ser. *Myriocarpis* et ser. *Anguria* plantis maturis sine cucurbitacino *D* vel flavonoidis *11*, *15* vel *19* (Brown et al., 1969), seminibusque ovatis absimilis.

Cucumis ser. *Metiluferi* consists of *C. metuliferus* and *C. rostratus*. The unique position of *Cucumis metuliferus* within the genus has been supported by flavonoid patterns (Brown et al., 1969), chloroplast DNA data (Perl-Treves, 1985; Perl-Treves and Galun, 1985), and various isozyme studies. *Cucumis rostratus* is placed here because of the similarity of its gross vegetative morphology with that of *C. metuliferus* and its geographical distribution. Some aspects of its fruit morphology are similar to those of *C. hystrix*. Biosystematic studies are needed to verify this placement or to establish the correct one.

25. **Cucumis metuliferus** E. Meyer ex Naudin, Ann. Sci. Nat. Bot., sér. 4, 11: 10. 1859.—TYPES (here designated): FRANCE. Muséum d'Histoire Naturelle, 1857, *C. Naudin s.n.* (lectotype: P); 1858, *C. Naudin s.n.* (syntype: P); 1859, *C. Naudin s.n.* (syntype: P).

    Cucumis tinneanus Kotschy & Peyritsch, Pl. tinn. 17, *t. 8*. 1867.—TYPE: SUDAN. Ad Wau et Meschra-Req prope originis fluminis Bahr-Ghasal, Jan 1864, *A.P.F. Tinne 47* (holotype: W).

Plants: Herbs; annual; with a woody rootstock; lacking tubers; monoecious. Stems procumbent or climbing; sulcate; not aculeate; pilose or setose; with nonbreakaway hairs. Stem hairs 1.9-3.7 mm long. Nodes not geniculate. Internodes (2.5-)5-10(-11.5) cm long. *Leaves*: Petioles (2-)4-7.5(-15) cm long; sulcate; not aculeate; pubescence a single type on each petiole; setose; with nonbreakaway hairs. Leaf blades trilobate, or 3- or 5-palmately lobed (weakly); with the margin serrate; ovate to broadly ovate, or shallowly ovate in outline; cordate at the base; with a basal sinus; (1-)1.5-3(-4) cm deep; acute to acuminate at the apex; (4-)6-12(-14) X (3.5-)5.5-10(-12) cm; 1.2-1.4 times longer than wide; hispid on the upper surface (many hairs broken off at base on older leaves); not aculeate on the veins below; hispid on the veins below; hispid on the intervenium below (many hairs broken off at base on older leaves); with nonbreakaway hairs. Central leaf-blade lobe entire; elliptic to ovate in outline; acute to acuminate at the apex; 2.5-5.5 X (1.5-)2.5-3.5 cm. Lateral leaf-blade lobes asymmetrical; entire; shallowly ovate to very shallowly ovate, or shallowly triangular in outline; broadly acute at the apex; 1-2.5 X 1.5-3.5 cm. *Tendrils*: Present; solitary; simple; 4-10.5 cm long; not aculeate; glabrate to hispid. *Inflorescences*: Unisexual or gynecandrous (rarely). *Bisexual inflorescences*: Racemose; ca 11-flowered; sessile. *Male inflorescences*: Fasciculate; 2-5-flowered; sessile. *Male flowers*: Pedicel terete in cross section; 7.2-10.4 mm long; hispid; without bracteoles. Hypanthium infundibular; 3.2-4 mm long; 2.9-4 mm in diam; hispid. Calyx lobes linear in outline; narrowly acute at the apex; 1.6-2.1 X 0.2-0.3 mm; sparsely hispid. Corolla infundibular; sparsely hispid outside; glabrous inside. Corolla tube ca 1.3 mm long; 4-5.5 mm in diam; sparsely hispid outside. Corolla lobes elliptic in outline; mucronate and obtuse at the apex; 5.1-8.3(-11) X 3.5-5.8(-9) mm; sparsely hispid outside; glabrous inside. Stamens separating from the hypanthium 1.2-2 mm from the base of the hypanthium. Filaments radially compressed in cross section; 0.4-1 mm long; ca 0.3 mm wide; glabrous. Anther thecae 2.2-2.6 mm long; puberulous. Anther connective oblong; unilobate; obtuse at apex; smooth at the apex; ca 0.9 X 0.6-0.7 mm. Disc cylindrical; 0.6-1.2 mm long; 0.4-1 mm in diam. *Female inflorescences*: A solitary flower. *Female flowers*: Pedicel terete in outline; ca 9 mm long; hispid; with nonbreakaway hairs; cylindrical. The lower 2/3 of hypanthium fused to the ovary. Fused portion of hypanthium ellipsoid; ca 5.6 mm long; ca 3.2 mm in diam; aculeate; glabrous. Aculei stiff; 2.6-2.8 mm long. Hyaline bristle at apex of aculeus ca 1.6 mm long; 1.3-1.6 times as long as the opaque base. The upper 1/3 of hypanthium free from the ovary. Free portion of hypanthium ca 3.2 mm long; ca 5.6 mm in diam; not aculeate outside; hispid outside; puberulent inside. Calyx lobes linear in outline; narrowly acute at the apex; 2-2.6 X 0.2-0.3 mm; sparsely hispid. Corolla sparsely hispid outside; sparsely puberulent inside. Corolla tube present; ca 1.6 mm long; ca 6.4 mm in diam; sparsely hispid outside. Corolla lobes broadly elliptic in outline; mucronate and obtuse at the apex; 6.4-9.7(-17) X 5.1-6(-12) mm; sparsely hispid outside; sparsely puberulent inside. Staminodes present; separating from the free portion of the hypanthium ca 1.6 mm above the ovary; 0.4-0.6 mm long; 0.2-0.4 mm in diam; puberulent. Style ca 1.2 mm long; ca 0.8 mm in diam; subtended by a circular disc. Disc ca 1 mm long; ca 2.2 mm in diam. Stigma ca 2.5 mm long; ca 4 mm in diam; lobate; with 6 finger-like projections on the margin. Finger-like stigmatic projections ca 2.4 mm long; 1.2-1.6 mm in diam; papillate. *Fruit*: Not geocarpic, maturing above ground and readily visible. Pedicel sulcate in cross section; 2-5(-7) cm long (lengthening as the fruit matures); hispid; with nonbreakaway hairs; cylindrical. Fruit monocolored; red or yellow; ellipsoid; 6-15 cm long; 3-6 cm in diam; aculeate; glabrous; blunt at the apex. Aculei terete; with hyaline bristle lost; 7-13 mm long; 1.9-4.5 mm in diam. *Seeds*: Ovate; 5.7-7.8 X 3.3-4 mm; 0.9-1 mm thick; unwinged. *Chromosome number*: 2n = 24 (Kozuchov, 1930; Deakin, Bohn & Whitaker, 1971; Dane & Tsuchiya, 1976; Ramachandran, 1984; Yadava, Singh & Arya, 1984; Nijs & Visser, 1985; Ramachandran & Narayan, 1985).

Distribution: Africa: Angola, Botswana, Ethiopia, Kenya, Malawi, Mozambique, Namibia, Senegal, South Africa, Sudan, Swaziland, Tanzania, Uganda, Zaire, Zambia, Zimbabwe, Cameroon, Central African Republic, Liberia, and Burkina. Southwest Asia: South Yemen and Yemen.

The name *Cucumis metuliferus* first appeared in *Reliquiae Schraderianae* in 1838 as a *nomen nudum* attributed to E. Meyer (Schrader, 1838). Naudin (1859) accepted *C. metuliferus* in his monograph of the genus and described it for the first time. He never saw any wild specimens, but it was grown for many years in the botanical garden at Paris. Therefore, his garden collections at P made in 1857, 1858, and 1859 are syntypes, and the 1857 collection, accompanied by his handwritten notes and watercolor drawing of the fruit, is chosen as lectotype.

Cucumis metuliferus is very easy to identify by its large, red fruits with a few large aculei. Meiotic and crossing studies (Raamsdonk et al., 1989) and chloroplast DNA (Perl-Treves and Galun, 1985) and isozyme analyses have all shown that *C. metuliferus* is isolated from the remainder of the genus. The chloroplast DNA studies also indicated similarities between *C. metuliferus* and *C. humifructus*, another species isolated within the genus.

26. **Cucumis rostratus** Kirkbride, sp. nov.—TYPES: NIGERIA. BENIN: Division Benin, Okomu Forest Reserve, compartment 69, 22 Jan 1948, *J.P.M. Brenan 8875* (holotype: K!; isotype: BOL!). IVORY COAST. Forêt de Divo, Oct 1959, *Aké-Assi 5495* (paratype: K!). Fig. 4.

A *Cucumere metulifero* floris feminei lobis 4-4.5 mm longis, floris feminei pedicello circa 19 mm longo, floris feminei aculei seta hyalina aculei basi opaca 4-9-plo longiora, fructu fusiformi eaculeatisque et *Cucumere hystrici* inflorescentiis maribus 1-3-floris, calycis lobis maribus 2-2.4 mm longis, floris feminei pedicello circa 19 mm longo, floris feminei aculei seta hyalina aculei basi opaca 4-9-plo longiora, calycis lobis femineis circa 3.7 mm longis, corollae femineae lobis circa 6 X 3.2 mm, fructus pedicello circa 2 cm longo, fructu fusiformi, circa 9 cm longo, circa 3.5 cm diametro, seminibus 5.8-6 X 2.6-2.7 X circa 1 mm absimilis.

Plants: Herbs; monoecious. Stems procumbent or climbing; sulcate; not aculeate; pilose; with nonbreakaway hairs. Stem hairs 1.2-2.8 mm long. Nodes not geniculate. Internodes 4-9 cm long. *Leaves*: Petioles 2-5.5 cm long; sulcate; not aculeate; pubescence a single type on each petiole; pilose; with nonbreakaway hairs. Leaf blades pentalobate; with the margin serrate; broadly ovate in outline; cordate at the base; with a basal sinus; 1-2.5 cm deep; acute at the apex; 5.5-9.5 X 4.5-8 cm; ca 1.2 times longer than wide; pilose on the upper surface; not aculeate on the veins below; pilose on the veins below; pilose on the intervenium below; with nonbreakaway hairs. Central leaf-blade lobe entire; narrowly triangular in outline; acute at the apex; 2.5-4 X 2.5-4.5 cm. Lateral leaf-blade lobes asymmetrical; entire; shallowly triangular in outline; acute to broadly acute at the apex; 0.5-1.5 X 1.5-3.5 cm. *Tendrils*: Present; solitary; simple; 2-6 cm long; not aculeate; pilose. *Inflorescences*: Unisexual. *Male inflorescences*: A solitary flower or fasciculate; 1-3-flowered; sessile. *Male flowers*: Pedicel terete in cross section; 4.8-8.5 mm long; pilose; without bracteoles. Hypanthium campanulate; ca 3.8 mm long; ca 2.5 mm in diam; pilose. Calyx lobes linear in outline; narrowly acute at the apex; 2-2.4 X ca 0.2 mm; hispid. Corolla infundibular; hispidulous outside; puberulent inside. Corolla tube 0.8-1 mm long; ca 3.5 mm in diam; hispidulous outside. Corolla lobes broadly elliptic in outline; obtuse and mucronate at the apex; 4-4.4 X ca 3.6 mm; hispidulous outside (pilose at the apex with 8-10 hairs); puberulent inside. Stamens separating from the hypanthium ca 1.6 mm from the base of the hypanthium. Filaments radially compressed in cross section; ca 0.6 mm long; ca 0.4 mm wide; glabrous. Anther thecae ca 2.2 mm long; puberulent. Anther connective

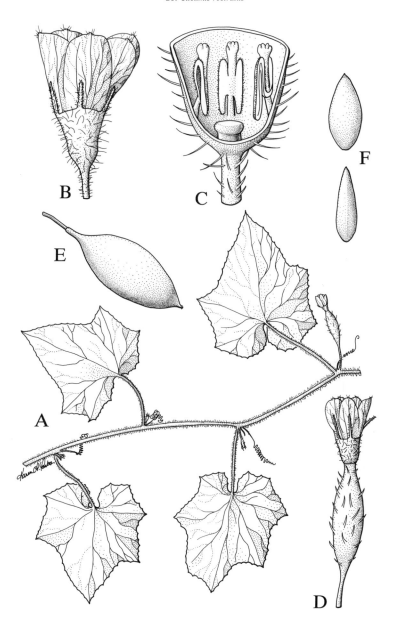

Fig. 4. *Cucumis rostratus*. A. Flowering stem with leaves, x ½. B. Male flower, x 5. C.
Male flower, opened to show stamens and disc, x 10. D. Female flower, x 2. E. Fruit, x ½.
F. Seed, face and side views, x 4½. (*Aké-Assi 5495* (K), E; *Brenan 8875* (K), A, B, C, D, F.)

obovate; unilobate; obtuse at apex; papillate at the apex; 0.8-1 X ca 0.6 mm. Disc cylindrical; ca 0.7 mm long; ca 1 mm in diam. *Female inflorescences*: A solitary flower. *Female flowers*: Pedicel sulcate in outline; ca 19 mm long; sparsely pilose; with nonbreakaway hairs; cylindrical. The lower 4/5 of hypanthium fused to the ovary. Fused portion of hypanthium ellipsoid; ca 16 mm long; ca 4 mm in diam; sparsely aculeate; glabrous. Aculei soft; 2-3 mm long. Hyaline bristle at apex of aculeus 1.8-2.4 mm long; 4-9 times as long as the opaque base. The upper 1/5 of hypanthium free from the ovary. Free portion of hypanthium ca 3.2 mm long; ca 2.2 mm in diam; not aculeate outside; sparsely short pilose outside. Calyx lobes linear in outline; narrowly acute at the apex; ca 3.7 X ca 0.2 mm; glabrate. Corolla puberulent outside. Corolla tube present; ca 0.9 mm long; ca 3.2 mm in diam; puberulent outside. Corolla lobes elliptic in outline; obtuse and mucronate at the apex; ca 6 X ca 3.2 mm; puberulent outside. *Fruit*: Not geocarpic, maturing above ground and readily visible. Pedicel sulcate in cross section; ca 2 cm long; pilose; with nonbreakaway hairs; cylindrical. Fruit spindle-shaped (strongly tapered at base and apex); ca 9 cm long; ca 3.5 cm in diam; not aculeate; glabrous; rostrate at the apex. Beak ca 0.4 cm long. *Seeds*: Seeds elliptic; 5.8-6 X 2.6-2.7 mm; ca 1 mm thick; unwinged.

　　Distribution: Africa: Ivory Coast and Nigeria.

　　Cucumis rostratus is an unusual member of the genus *Cucumis*. According to its collectors' annotations, it is found in forested environments including "high [tall] forests" (*Brenan 8875*). In general, the genus *Cucumis* is found in drier, more open habitats. The fruit exhibits characteristics found in no other member of the genus. They are spindle-shaped, tapering strongly at both ends. Thus the base of the fruit imperceptibly unites with its pedicel, and the apex tapers into a beak ca 0.4 cm long. The specific epithet, *rostratus* emphasizes the latter character.

Cucumis sect. Melo

Cucumis sect. **Melo** (Miller) Kirkbride, comb. & stat. nov.—BASIONYM: *Melo* Miller., Gard.
　　　Dict. abr. ed. 4 [without pagination]. 1754.—TYPE: *Melo cantalupa* (Naudin)
　　　Pangalo (lectotype, = *Cucumis melo* Linnaeus). Species 27-30.

　　Within subgen. *Melo*, this section brings together those series without aculeate fruit, ser. *Hirsuti*, ser. *Humifructosi*, and ser. *Melo*. It is a convenient morphological grouping that corresponds, in its composition, to the first lead of couplet 4 in the Key to the Species of *Cucumis* and to the major divisions of Naudin (1859) and Cogniaux (1881).

Cucumis ser. **Hirsuti** Kirkbride, ser. nov.—TYPE: *Cucumis hirsutus* Sonder in Harvey &
　　　Sonder. Species 27.

　　A ser. *Melone* plantis dioecis, inflorescentiis maribus pedunculatis, filamentisque ad basim puberulis apicem versus glabris differt.

　　Cucumis ser. *Hirsuti* is well set off from the remainder of the genus by its unique combination of morphological characters, see couplet 4 of the Key to Subgenera, Sections, and Series of *Cucumis*, and by its distinctive flavonoid pattern with just two compounds (Brown et al., 1969). For such a common species, it is singularly lacking in biosystematic information.

27. **Cucumis hirsutus** Sonder in Harvey & Sonder, Fl. cap. 2: 497. 1862.—TYPES: SOUTH AFRICA. Inter Wonderfontyn et Moojerivier, *C.L.P. Zeyher 581* (lectotype: MEL!; isolectotypes: BM! K! P! S!); Mooje River, *J. Burke 297* (syntype: K!); Port Natal, *C.F.F. Krauss 91* (syntypes: BM! PRE!).
Cucumis hirsutus var. *major* Cogniaux, Monogr. phan. 3: 490. 1881.—TYPE: MOZAMBIQUE. MORAMBALLA: 14°-19° lat. S., *J. Kirk s.n.* (holotype: K; isotype: MEL).
Cucumis sonderi Cogniaux, Monogr. phan. 3: 489. 1881.—TYPE: SOUTH AFRICA. NATAL: Port Natal, *W. Gueinzius 397* (lectotype: W!); *W.T. Gerrard 658* (syntype: K); *W.B.L. Grant s.n.* (syntype: K).
Cucumis welwitschii Cogniaux, Monogr. phan. 3: 490. 1881.—TYPE: ANGOLA. Inter flum. Lucala et locum Zamba, distr. Ambaca, *F.M.J. Welwitsch 803* (lectotype: BM!; isolectotype: BM! BR! LISU!).—*Cucumis hirsutus* var. *welwitschii* (Cogniaux) R. Fernandes & A. Fernandes, Mem. Junta Invest. Ultram., ser. 2, 34: 86. 1962.
Cucumis hirsutus var. *dissectus* Cogniaux, Bull. Herb. Boissier 3: 418. 1895.—TYPE: SOUTH AFRICA. Transvaal: Houtbosch, *A. Rehmann 6310* (holotype: BR!).
Cucumis seretii de Wildeman, Ann. Mus. Congo, sér. 5, Bot., 3: 143. 1909.—TYPE: ZAIRE. Savane nord-orientale, Mar 1907, *F. Seret 811* (holotype: BR!).
Cucumis homblei de Wildeman ex Cogniaux, Bull. Jard. Bot. Etat 5: 100. 1916.—TYPES: ZAIRE. KATANGA: Elisabethville, Mar 1912, *H.A. Homble 227* (lectotype: BR!); Lupaka R., 6 Feb 1908, *T. Kassner 2463* (syntypes: BM! BR!).
Cucumis wildemanianus Cogniaux, Bull. Jard. Bot. Etat 5: 111. 1916.—TYPE: ZAIRE. KATANGA: Bord de la rivière Kapiri, Feb 1913, *H.A. Homble 1200* (holotype: BR!).
Cucumis hirsutus var. *ovatus* Cogniaux in Cogniaux & Harms, Pflanzenr. 88: 134. 1924.—TYPE. SOUTH AFRICA. Natal: Highland Station, 19 Mar 1894, *O. Kuntze s.n.* (holotype: BR!).
Cucumis gossweileri Norman in Exell, Good & Norman, J. Bot. 67: 195. 1929.—TYPE: ANGOLA. Cubango: East of Forte Princessa Amelia, 6 Jan 1907, *J. Gossweiler 2501* (holotype: BM!).
Cucumis seretioides Süssengüth in Süssengüth & Merxmüller, Trans. Rhodesia Sci. Assoc. 43: 135. 1951.—TYPE. ZIMBABWE: Marandellas, 30 Jan 1942, *G. Dehn 582* (holotype: M!; isotypes: K! SRGH!).

Plants: Herbs; perennial; with a woody rootstock; lacking tubers; dioecious. Stems procumbent (or rarely scandent); sulcate; not aculeate; hirsute to pilose; with nonbreakaway hairs. Stem hairs 0.8-3.2 mm long. Nodes not geniculate. Internodes 2.5-4 cm long. *Leaves*: Petioles 0.5-4(-5.5) cm long; terete or sulcate in cross section; not aculeate; pubescence a single type on each petiole; hirsute to pilose; with nonbreakaway hairs. Leaf blades 3- or 5-palmately lobed, or entire; with the margin serrate (weakly so, with a vein terminating in the apex of each tooth); ovate to broadly ovate, narrowly ovate (rarely), or elliptic (very rarely) in outline; subcordate to cordate, subtruncate, or obtuse at the base; with or without a basal sinus; 0-1.8 cm deep; narrowly acute to broadly acute, or obtuse at the apex; (2.5-)4-14(-21) X (1.5-)2.5-7(-10.5) cm; (1.4-)1.6-3.2(-4.3) times longer than wide; pilose on the upper surface; not aculeate on the veins below; pilose on the veins below; pilose on the intervenium below; with nonbreakaway hairs. Central leaf-blade lobe entire; broadly ovate or narrowly oblong in outline; acute to broadly acute at the apex; 3.5-4 X (0.5-)3-3.5 cm. Lateral leaf-blade lobes asymmetrical; entire; shallowly triangular, narrowly oblong to broadly oblong, or square (rarely) in outline; broadly acute at the apex; 0.5-1 X (0.5-)1-1.5 cm. *Tendrils*: Present; solitary; simple; 2-5 cm long; not aculeate; hirsute (with the apex glabrate). *Inflorescences*: Unisexual. *Male inflorescences*: Reduced racemose or modified compound dichasial (with scorpioid cymose branches, frequently with a coaxillary flower); 1- (rarely) or 2-11-flowered; pedunculate. Peduncle 1-13 or 30-45 cm long. *Male flowers*: Pedicel terete in cross section; 8-85 mm long; pilose; without bracteoles. Hypanthium campanulate; 4.8-8.8 mm long; 3.2-6.4 mm in diam; pilose (to densely so). Calyx lobes narrowly triangular or linear in outline; narrowly acute at the apex; (0.8-)2.4-6.1 X (0.3-)0.5-0.8 mm; pilose (to sparsely so). Corolla campanulate; hirsute (to sparsely so, with the hairs on the veins slightly longer) outside; glabrous inside. Corolla tube 0.4-2 or 6-10 mm long; 2.5-4.8 or 8-10 mm in diam; hirsute outside. Corolla lobes ovate or broadly triangular in outline; narrowly acute to broadly acute at the apex; (2-)4.8-8.8 X (1.2-)2.8-3.8 or 5.6-6.4 mm; hirsute outside; glabrous inside. Stamens separating from the hypanthium 1-3.2 mm from the base of the hypanthium. Filaments terete in cross section; ca 1 mm long; 0.2-0.3 mm wide; with basal 1/2-2/3 puberulent and glabrous apically. Anther thecae 1.9-2.2 mm long; puberulous. Anther connective ovate (to narrowly so); unilobate; obtuse at apex; papillate at the apex; ca 0.8 X 0.3-0.8 mm. Disc cylindrical; 0.4-0.8 mm long; 0.3-0.8 mm in diam. *Female inflorescences*: A solitary flower or fascicles; 1-3-flowered. *Female flowers*: Pedicel terete in outline; 6-20(-70) mm long; pilose; with nonbreakaway hairs; cylindrical. The lower 2/3 of hypanthium fused to the ovary. Fused portion of hypanthium ovoid (or narrowly so); 5.6-9 mm long; 2.8-4.4 mm in diam; not aculeate; densely pilose. The upper 1/3 of hypanthium free from the ovary. Free portion of hypanthium 2.8-4.4 mm long; 2.4-5.3 mm in diam; not aculeate outside; pilose outside; glabrous inside. Calyx lobes narrowly triangular in outline; narrowly acute at the apex; 1-5.9 X 0.2-0.6(-1.2) mm; puberulent or pilose. Corolla puberulent outside (to sparsely so); glabrous inside. Corolla tube present; 1-4(-11) mm long; 3-6 mm in diam; puberulent outside (to sparsely so). Corolla lobes elliptic to broadly elliptic in outline; acute to broadly acute at the apex; 9.6-14.4 X 4.8-11.2 mm; glabrous (to sparsely so) outside; glabrous inside. Staminodes present; separating from the free portion of the hypanthium 0.8-1.6 mm above the ovary; 1-2 mm long; 0.2-0.4 mm in diam; sparsely puberulent. Style 1.2-2.4 mm long; 0.4-0.8 mm in diam; subtended by a circular disc. Disc ca 0.4 mm long; 1.2-2 mm in diam. Stigma 2.2-3.6 mm long; 1.6-3.6 mm in diam; entire to sublobate. Finger-like stigmatic projections smooth or papillate. *Fruit*: Not geocarpic, maturing above ground and readily visible. Pedicel sulcate in cross section; (2-)3-5(-9.5) cm long; hirsute to pilose; with nonbreakaway hairs; cylindrical. Fruit bicolored with longitudinal stripes from base to apex; red or brown; with dark green longitudinal stripes; broadly ellipsoid to globose; 3-5 cm long; 3-4.5 cm in diam; not aculeate; when immature pilose and when mature glabrous; blunt at the apex. *Seeds*: Ovate; 5.6-8.3 X 4-5.6 mm; 2-5 mm thick; unwinged. *Chromosome number*: 2n = 24 (Shimotsuma, 1965).

    Distribution: Africa: Angola, Botswana, Burundi, Congo, Kenya, Malawi, Mozambique, South Africa, Sudan, Swaziland, Tanzania, Zaire, Zambia, and Zimbabwe.

    The great number of herbarium collections available show that *C. hirsutus* is a widespread and frequent species in southern and eastern Africa. The density, length, and distribution of pubescence on its organs are extremely but continuously variable and not geographically related. If pubescence characters were used, it could be divided into a number of taxa without any significant separation between them. It can be easily distinguished from the remainder of the genus by the morphological characters set forth in couplet 4 of the Key to the Subgenera, Sections, and Series of *Cucumis* and couplet 9 of the Key to the Species of *Cucumis*.

Cucumis ser. **Humifructosi** Kirkbride, ser. nov.—TYPE: *Cucumis humifructus* Stent. Species 28.

A ser. *Hirsutis* et ser. *Melone* caudicibus lignosis nullis, cirrhis axillaribus 5-8 ad nodum fasciculatis, inflorescentiis androgynis, disco ad styli basim nullo, fructibusque geocarpicis differt.

Cucumis ser. *Humifructosi*, with the single species *C. humifructus*, occupies an isolated position within the genus. It is the only *Cucumis* with geocarpic fruit; for other unique morphological characters see couplet 3 of the Key to the Subgenera, Sections, and Series of *Cucumis*. In addition, it has a unique flavonoid compound (Brown, 1969); in isozyme studies (Dane, 1976, 1983; Esquinas-Alcazar, 1977, Perl-Treves and Galun, 1985) it was associated with some collections of *C. sagittatus* and well separated from the remainder of the genus; and, based on chloroplast DNA mutations, it was separated from the bulk of *Cucumis* and associated with *C. metuliferus* with 3 mutations in common between them and 5 mutations separating them.

28. **Cucumis humifructus** Stent, Bothalia 2(1b): 356. 1927.—TYPES: SOUTH AFRICA. Transvaal: Pretoria, grown in the grounds of Botanical Laboratory from seed sent by Mr. E. E. Galpin, Naboomspruit, Waterberg (*Galpin M. 719*), 13 Feb 1925, *S.M. Stent s.n.* (holotype: PRE!); Waterberg, Naboomspruit, Farm Roodepoort, 19 Feb 1919, *E.E. Galpin M. 146* (paratypes: P! PRE!); grown in Mosdene Garden, 23 Jan 1925, *E.E. Galpin M. 146* (paratypes: BM! P! PRE!).

Plants: Herbs; annual; without a woody rootstock; lacking tubers; monoecious. Stems procumbent; sulcate; not aculeate; villous (to sparsely so); with nonbreakaway hairs. Stem hairs 1.5-2.5 mm long. Nodes not geniculate. Internodes 4-18 cm long. *Leaves*: Petioles 2-7(-10) cm long; sulcate; not aculeate; pubescence a single type on each petiole; villous (to sparsely so, with the hairs simple); with nonbreakaway hairs. Leaf blades weakly 5-palmately lobed; with the margin weakly serrate (with a vein terminating in the apex of each tooth); broadly ovate in outline (to very broadly so); broadly cordate at the base; with a basal sinus; 0.5-3 cm deep; abruptly and shortly acuminate at the apex; 5.5-12(-18) X 5.5-11.5(-17) cm; 1-1.1 times longer than wide; pilose on the upper surface; not aculeate on the veins below; pilose on the veins below; pilose on the intervenium below; with nonbreakaway hairs. Central leaf-blade lobe entire; shallowly ovate or very shallowly ovate (rarely) in outline; abruptly and shortly acuminate at the apex; 2-3(-4) X 3.5-6.5(-12) cm. Lateral leaf-blade lobes asymmetrical; entire; very shallowly ovate in outline; broadly acute to obtuse at the apex; 0.5-1(-2.5) X 1.5-4(-9) cm. *Tendrils*: Present; 5-8 at a node in an axillary fascicle; simple; 1-5 cm long; not aculeate; glabrous. *Inflorescences*: Androgynous. *Bisexual inflorescences*: Racemose (the basal flower feminine and the remainder masculine, with the female flower opening first or rarely simultaneously with the male flowers); 8-14-flowered; sessile. *Male flowers*: Pedicel terete in cross section; 8-44 mm long; glabrous (with a zone of disarticulation at the apex); without bracteoles. Hypanthium campanulate; ca 0.5 mm long; 2.5-3 mm in diam; pilose (to sparsely so). Calyx lobes linear in outline; narrowly acute at the apex; 1.5-2 X ca 0.2 mm; pilose. Corolla infundibular; puberulent outside; puberulent inside. Corolla tube 1-1.5 mm long; 2-2.5 mm in diam; puberulent (to sparsely so) outside. Corolla lobes ovate in outline; mucronate at the apex (with the mucro ca 0.1 mm long); 4.8-6.8 X 3.2-4 mm; puberulent (to sparsely so) outside; puberulent (to sparsely so) inside. Stamens separating from the hypanthium ca 1.2 mm from the base of the hypanthium. Filaments terete in cross section; ca 0.4 mm long; 0.1-0.2 mm wide; glabrous. Anther thecae 1.6-2 mm long; puberulent. Anther connective ovate; unilobate; obtuse at apex; smooth at the apex; 0.6-0.8 X ca 0.4 mm. Disc cylindrical; ca 0.4 mm long; ca

0.8 mm in diam. *Female flowers*: Pedicel terete in outline; ca 4 mm long; densely pilose; with nonbreakaway hairs; cylindrical. The lower 2/3 of hypanthium fused to the ovary. Fused portion of hypanthium narrowly ovoid; ca 4.8 mm long; ca 1.6 mm in diam; not aculeate; retrorse sericeous. The upper 1/3 of hypanthium free from the ovary. Free portion of hypanthium ca 2.8 mm long; ca 2.8 mm in diam; not aculeate outside; pilose outside; puberulent inside. Calyx lobes narrowly oblong to linear in outline; narrowly acute at the apex; 1.5-2.5 X ca 0.2 mm; pilose (to sparsely so). Corolla puberulent outside; puberulent inside. Corolla tube present; 0.8-2 mm long; ca 3 mm in diam; puberulent (to sparsely so) outside. Corolla lobes broadly ovate in outline; mucronate at the apex (with the mucro 0.2-0.3 mm long); ca 3.2 X ca 1.2 mm; puberulent (to sparsely so) outside; puberulent (to sparsely so) inside. Staminodes present; separating from the free portion of the hypanthium ca 2.8 mm above the ovary; 0.2-0.4 mm long; ca 0.6 mm in diam; glabrous. Style ca 0.8 mm long; ca 0.6 mm in diam; not subtended by a disc. Stigma ca 2.5 mm long; ca 2 mm in diam; lobate; with 3 finger-like projections on the margin. Finger-like stigmatic projections 0.4-0.6 mm long; 0.2-0.4 mm in diam; smooth. *Fruit*: Geocarpic, maturing below ground and not visible. Pedicel sulcate in cross section; 15-25 cm long; glabrous; with nonbreakaway hairs; cylindrical. Fruit monocolored; dull white or yellow; globose to ellipsoid (broadly), or globose to ovoid (broadly); 2.5-5 cm long; 2.5-5 cm in diam; not aculeate; glabrous; blunt at the apex. *Seeds*: Elliptic; 15-20 X 6-9 mm; ca 2 mm thick; unwinged. *Chromosome number*: 2n = 24 (Ramachandran, 1984; Ramachandran & Narayan, 1985).

   Distribution: Africa: Angola, Ethiopia (very rarely), Kenya, Namibia, South Africa, Zaire, Zambia, and Zimbabwe.

   In South Africa, *Cucumis humifructus* always seemed to occur near old holes of the aardvark or antbear, *Orycteropus afer* (Pallas), and local farmers observed that under the center of each plant were the remains of antbear dung which the antbear had buried. Apparently the antbear eats the fruits of *C. humifructus*, buries its dung, and plants the undamaged seeds. Thus, the seeds of *C. humifructus* are dispersed, planted, and fertilized by the antbear. Because the pulp of the fruits is soft and watery, probably it is a source of water for the antbear during the dry winter season (Meeuse, 1955, 1958). Similar observations on the antbear and *C. humifructus* have been made in the central uplands of Zaire (Verheyen, 1951). Possibly there is an interdependent relationship between *C. humifructus* and the antbear throughout their ranges of distribution.

Cucumis ser. **Melo** (Miller) Kirkbride, comb. & stat. nov.—BASIONYM: *Melo* Miller, Gard.
   Dict. abr. ed. 4 [without pagination]. 1754.—TYPE: *Melo cantalupa* (Naudin)
   Pangalo (lectotype, = *Cucumis melo* Linnaeus). Species 29 and 30.
   Melo sect. *Eumelon* Pangalo, Bot. Zhurn. (Moscow & Leningrad) 35(6): 574. 1950.
   Melo ser. *Graciles* Pangalo, Bot. Zhurn. (Moscow & Leningrad) 35(6): 575. 1950.
   Melo ser. *Rigidi* Pangalo, Bot. Zhurn. (Moscow & Leningrad) 35(6): 577. 1950.
   Melo sect. *Melonoides* Pangalo, Bot. Zhurn. (Moscow & Leningrad) 35(6): 580. 1950.
   Melo sect. *Bubalion* Pangalo, Bot. Zhurn. (Moscow & Leningrad) 35(6): 580. 1950.
   Melo sect. *Archimelon* Pangalo, nom. nud., Bot. Zhurn. (Moscow & Leningrad) 35(6):
   574. 1950.

   Isozyme data (Dane, 1976, 1983; Esquinas-Alcazar, 1977; Perl-Treves and Galin, 1985) separated this group from the remainder of the genus, but either associated it with *C. metuliferus* (Dane, 1976, 1983; Esquinas-Alcazar, 1977) or *C. humifructus* (Perl-Treves and Galun, 1985). The chloroplast DNA study (Perl-Treves, 1985; Perl-Treves and Galun, 1985) distinctly separated *C. melo* and *C. sagittatus* from the rest of the genus but near to each other, with just two mutations separating the two species. Fruit set is stimulated in *C. melo* and *C. sagittatus* by members of *Cucumis* ser. *Angurioidei*, but not by either *C. humifructus* or *C. metuliferus* (Raamsdonk et al., 1989). Cucurbitacin F was detected (Enslin and Rehm, 1957) only in *C. sagittatus*.

29. **Cucumis melo** Linnaeus, Sp. pl. *ed. 1*, 1011. 1753.

All synonyms are listed in Appendix 1.

Plants: Herbs or vines; annual or perennial; with or without a woody rootstock; lacking tubers; monoecious or andromonoecious (rarely in cultivated individuals). Stems procumbent; sulcate; not aculeate; hispid, retrorse-strigose, or lanate (rarely); with nonbreakaway hairs. Stem hairs 0.5-3.6 mm long. Nodes not geniculate. Internodes 1.5-20 cm long. *Leaves*: Petioles 1.5-20 cm long; sulcate; not aculeate; pubescence a single type on each petiole; hispid (sometimes sparsely so), hispidulous, retrorse-strigose, or lanate (rarely); with nonbreakaway hairs. Leaf blades entire, trilobate, pentalobate, or 3- or 5-palmately lobed (Australian plants show the deepest lobation); with the margin entire or serrate (to weakly so); broadly ovate or very broadly ovate in outline; cordate or acute (rarely in some Australian plants) at the base; with or without (rarely in some Australian plants) a basal sinus; 0.5-2.5(-5) cm deep; acute, broadly acute, or obtuse at the apex; 2-14(-26) X 2-15(-26) cm; 0.8-1.3 times longer than wide; hispid, hispidulous, or pilose on the upper surface; not aculeate on the veins below; hispid, hispidulous, or pilose on the veins below; hispid, hispidulous, or pilose on the intervenium below; with nonbreakaway hairs. Central leaf-blade lobe entire or pinnatifid (rarely in some Australian plants); broadly ovate, shallowly, narrowly oblong, or elliptic (rarely) in outline; obtuse, acute, or broadly acute at the apex; (0.5-)1.2-6(-8) X 1-9 cm. Lateral leaf-blade lobes asymmetrical; entire or pinnatifid (rarely in some Australian plants); shallowly ovate, very shallowly ovate, broadly ovate, or narrowly oblong in outline; broadly acute, obtuse, or narrowly acute (rarely) at the apex; 0.5-3.5 X 1-7 cm. *Tendrils*: Present; solitary; simple; 2-14 cm long; not aculeate; hispid (sparsely), hispidulous (sparsely), or retrorse-strigose. *Inflorescences*: Unisexual. *Male inflorescences*: Fasciculate or paniculate (rarely, with 2 or 3 short scorpioid cymous branches); 2-7(-18)-flowered; sessile. *Male flowers*: Pedicel terete in cross section; 2.4-15(-55) mm long; hispidulous or retrorse-strigose; without bracteoles. Hypanthium campanulate or infundibular; 2.8-4(-5.6) mm long; 2-3.2(-5.6) mm in diam; pilose or hispidulous. Calyx lobes linear or narrowly triangular in outline (rarely in the largest cultivated plants); narrowly acute at the apex; 1.2-3.6 X 0.2-0.4 mm; pilose to glabrate, hispidulous (on the veins), or hispid. Corolla infundibular; puberulent, pilose outside, or (rarely in some Australian plants) glabrous outside; puberulent (or minutely so) or glabrous inside. Corolla tube 0.8-2 mm long; 3.2-4(-11) mm in diam; pilose, puberulent, or glabrous (rarely in some Australian plants) outside. Corolla lobes elliptic, broadly elliptic, or broadly ovate in outline; obtuse or acute, and mucronate at the apex; 2-9(-24) X 2-5(-20) mm; puberulent, pilose, or glabrous (rarely in some Australian plants) outside; puberulent (or minutely so) or glabrous inside. Stamens separating from the hypanthium (0.8-)1.2-1.6 mm from the base of the hypanthium. Filaments terete or radially compressed (rarely in some Australian plants) in cross section; 0.3-0.8 mm long; 0.2-0.6 mm wide; glabrous. Anther thecae 0.9-2(-2.8) mm long; puberulent (on edges). Anther connective obovate or elliptic (rarely in cultivated plants); unilobate or bilobate; obtuse or acute at apex; smooth or papillate at the apex; 0.4-1.6 X 0.4-1.2 mm. Disc cylindrical or consisting of 3 papillae (rarely in very large cultivated plants); 0.3-1.4 mm long; 0.6-2.4 mm in diam. *Female inflorescences*: A solitary flower. *Female flowers*: Pedicel sulcate in outline; 1.6-12(-30) mm long; hispidulous or retrorse-strigose (rarely in some Australian plants); with nonbreakaway hairs; cylindrical. The lower 2/3 of hypanthium fused to the ovary. Fused portion of hypanthium ellipsoid; 4-11(-14) mm long; 2-5 mm in diam; not aculeate; pilose, lanate, retrorse sericeous, or antrorse sericeous (with spreading or appressed hairs). The upper 1/3 of hypanthium free from the ovary. Free portion of hypanthium 2.4-4.5 mm long; 1.6-4.8 mm in diam; not aculeate outside; pilose or hispidulous outside (with spreading hairs); puberulent inside (minutely) or glabrous inside. Calyx lobes linear, triangular, or narrowly elliptic in outline; narrowly acute at the apex; 1.6-2.8(-8) X 0.2-0.8(-1.6) mm; glabrate, pilose (sparsely), or puberulent. Corolla puberulent or pilose (sparsely, rarely) outside; puberulent (or minutely so) or glabrous inside. Corolla tube present or absent (rarely); 0.8-1.6(-2.8) mm long; 3.6-5 mm in diam; puberulent or antrorse-strigose (sparsely) outside. Corolla lobes broadly obovate, elliptic to broadly elliptic, or ovate in outline; obtuse and mucronate at the apex; 3.6-9.2(-20) X 3.2-6.4(-17) mm; puberulent outside; puberulent (to minutely so) or glabrous (rarely) inside. Staminodes present (perfect flowers differing in the lack of staminodes and presence of stamens); separating from the free portion of the hypanthium 0.8-2 mm above the ovary; 0.2-0.4(-1.1) mm long; 0.1-0.3(-1) mm in diam; glabrous. Style 0.8-1.6(-2.4) mm long; 0.4-0.6(-1.1) mm in diam; subtended by a circular disc. Disc 0.4-1 mm long; (0.8-)1.4-1.6(-2.9) mm in diam. Stigma 1.4-2.4(-4.8) mm long; 1.4-2.2(-4.8) mm in diam; lobate; with 3, 6, or 5 finger-like projections on the margin. Finger-like stigmatic projections 0.8-1.8(-3.2) mm long; 0.4-1(-2.4) mm in diam; smooth or papillate. *Fruit*: Not geocarpic, maturing above ground and readily visible. Pedicel sulcate in cross section; 1.5-4 cm long; hispidulous (to sparsely so), glabrous, or retrorse-strigose (rarely in some Australian plants); with nonbreakaway hairs; cylindrical. Fruit monocolored or bicolored with longitudinal stripes from base to apex; green, red, yellow, white, or brown; with light green, yellow, brown, white, or dark green longitudinal stripes; ellipsoid, globose, cylindrical, ovoid, or obovoid; 2-12(-100) cm long; 2-5(-20) cm in diam; not aculeate; glabrous or pilose (sparsely so); blunt at the apex. *Seeds*: Ovate or elliptic; 4-8(-18) X 2.5-4(-13) mm; 1-2 mm thick; unwinged or apically winged (rarely in some Australian plants [*Must 1507*]). Wing transversely oblong; truncate at apex; ca 1.6 X 1.8-2 mm. *Chromosome number*: 2n = 24 (Kozuchov, 1930; Bhaduri & Bose, 1947; Naithani, 1947; Chandola, Bhatnagar, & Tokuta, 1965; Shimotsuma, 1965; Deakin, Bohn & Whitaker, 1971; Ramachandran, 1984; Singh & Yadava, 1984; Yadava, Singh & Arya, 1984; Ramachandran & Narayan, 1985), 22 (rarely, Chandola, Bhatnagar, & Tokuta, 1965), or 20 (rarely, Chandola, Bhatnagar, & Tokuta, 1965).

   Distribution: Africa: Angola, Benin, Burkina, Cameroon, Cape Verde Islands, Central African Republic, Chad, Egypt, Ethiopia, Gambia, Ghana, Guinea-Bissau, Ivory Coast, Kenya, Madagascar, Malawi, Maldive Islands, Mali, Mauritania, Mozambique, Niger, Nigeria, Senegal, Seychelles, Sierra Leone, Somalia, South Africa, Sudan, Tanzania, Uganda, Zaire, Zambia, and Zimbabwe. Southwest Asia: Iran, Iraq, Oman, Saudi Arabia, South Yemen, and Yemen. Asia: Afghanistan, Bangladesh, Burma, China, India, Japan, Korea, Nepal, Pakistan, Sri Lanka, and Thailand. Malesia: Indonesia, New Guinea, and Philippines. Australia. Pacific: Fiji Islands, Guam, New Britain, Papua New Guinea, Samoa, Solomon Islands, and Tonga Islands. Introduced: widely cultivated all over the world.

Key to the Subspecies of *Cucumis melo*

1. Fused portion of hypanthium pilose or lanate (with spreading hairs, observable on the
      youngest fruits)   *C. melo* subsp. *melo*
1. Fused portion of female hypanthium retrorse sericeous or antrorse sericeous (with appressed
      hairs, observable on the youngest fruits)   *C. melo* subsp. *agrestis*

29a. **Cucumis melo** Linnaeus subsp. **melo**—TYPE: SWEDEN. Plant cultivated at Uppsala, Sweden (lectotype: LINN, sheet number 1152.8! [Meeuse, 1962]).

Female flowers: Fused portion of hypanthium pilose or lanate (with spreading hairs, observable on the youngest fruits).

Distribution: Africa: Angola, Cameroon, Cape Verde Islands, Central African Republic, Egypt, Ethiopia, Ghana, Guinea-Bissau, Kenya, Maldive Islands, Mali, Niger, Nigeria, South Africa, Sudan, Tanzania, and Zambia. Southwest Asia: Iran. Asia: Afghanistan, Burma, China, India, Japan, and Pakistan. Malesia: New Guinea. Australia. Pacific: Fiji Islands and Papua New Guinea. Introduced: widely cultivated all over the world.

29b. **Cucumis melo** Linnaeus subsp. **agrestis** (Naudin) Pangalo in Zhukovsky, La Turquie agricole 534. 1933.—BASIONYM: *Cucumis melo* Linnaeus var. *agrestis* Naudin.

Female flowers: Fused portion of hypanthium retrorse sericeous or antrorse sericeous (with appressed hairs, observable on the youngest fruits).

Distribution: Africa: Benin, Chad, Ethiopia, Ghana, Ivory Coast, Malawi, Maldive Islands, Mozambique, Niger, Senegal, Seychelles, Somalia, South Africa, Sudan, Tanzania, Uganda, and Zimbabwe. Southwest Asia: Iran, Saudi Arabia, and Yemen. Asia: Burma, China, India, Japan, Korea, Nepal, Pakistan, Sri Lanka, and Thailand. Malesia: Indonesia, New Guinea, and Philippines. Australia. Pacific: Guam, New Britain, Papua New Guinea, Samoa, Solomon Islands, and Tonga Islands. Introduced: cultivated in India and eastern Asia, widely adventive as a weed in tropical areas.

There has been some confusion surrounding the lectotype of *C. melo*. Meeuse (1962) selected sheet number 1152.8 at LINN. Then Jeffrey (1967) selected a specimen at BM cultivated in Clifford's garden at Hartekamp. In 1980, Jeffrey selected the phrase name, *Cucumis foliorum angulis rotundis, pomis torulosis*, in *Hortus cliffortianus* (Linnaeus, 1738) as lectotype, commenting, "...in spite of the statement by Jeffrey (1967) to the contrary, no specimen from Hartekamp is now to be found in the Clifford Herbarium (BM); in its absence, the phrase-name itself is to be made the lectotype." Sheet number 1152.8 was in LINN when *Species plantarum* (Linnaeus, 1753) was published (C.E. Jarvis, pers. comm.); it is a member of *C. melo* subsp. *melo*; it does not conflict with the protologue of *C. melo*, so, Meeuse's lectotype is cited here as required by article 8 of the ICBN, Berlin (Greuter et al., 1988).

Cucumis melo is the most variable species of the genus *Cucumis*. The variation of the fruits surpasses that found in all the rest of the genus. They vary in size, internal color, and surface ornamentation and color. Because of this tremendous man-induced and -maintained variability of the fruit, the infraspecific classification of *C. melo* based on fruit characters has been a long-standing topic of interest (Naudin, 1859; Alfeld, 1866; Cogniaux, 1881; Harz, 1885; Cogniaux and Harms, 1924; Hassib, 1938; Pangalo, 1950; Filov, 1960; Chakravarty, 1966; Mallick and Masui, 1986; Perrino et al., 1986; Munger and Robinson, 1991). Naudin (1859) proposed the first practical scheme of infraspecific taxa within the Linnean hierarchy with 10 varieties. Various schemes have been proposed through intercalation of ranks, use of additional ranks to express cultivar biology, change of ranks, etc., but Naudin's system is still the most reasonable one (Munger and Robinson, 1991). These cultivar groups should be treated under the code of nomenclature for cultivated plants (Brickell, et al., 1980) and not under the code of nomenclature for botanical names (Greuter, et al., 1988). Consequently, I have chosen to use the character of pubescence type on the female-flower hypanthium to delimit the botanical subspecies proposed here for *C. melo*.

Australia is a center of complex variation for *C. melo* (I.R. Telford, pers. comm.). Forms have developed there with highly dissected leaves and unusual pubescence on the female-flower hypanthium. Study of herbarium specimens reveals no fixed pattern of variation. The characters are found in various combinations. I have chosen to treat this variation under the two subspecies based upon the pubescence types. Further biosystematic investigation is needed to understand the variation in Australian *C. melo* and to open up a new pool of genetic variation potentially useful for agriculture.

30. **Cucumis sagittatus** Peyritsch in Wawra & Peyritsch, Sitzungsber. Kaiserl. Akad. Wiss. Math.-Naturwiss. Cl. 38: 567. 1860.—TYPE (here designated): ANGOLA. Prope Benguelam, *H. Wawra 263* (lectotype: W!; isolectotypes: L! W!).

Cucumis angolensis Hooker filius ex Cogniaux, Monogr. phan. 3: 487. 1881.—TYPES: ANGOLA. Distr. Mossamedes: Ad ripas flum. Bero (Charneca da boca do Bero), Jul 1859, *F.M.J. Welwitsch 831* (holotype: BR!; isotypes: BR! [fragments] LISU!).

Cucumis dinteri Cogniaux, Bull. Herb. Boissier, sér. 2, 1: 882. 1901.—TYPE: NAMIBIA. Hereroland, 12 Feb 1900, *K. Dinter 1436* (holotype: BR!; isotype: Z).

Cucumis cogniauxianus Dinter ex Cogniaux & Harms, Pflanzenr. 88: 131. 1924.—TYPES: NAMIBIA. Hereroland: Aus, *K. Dinter 1115* (syntype: not seen); Satansplatz, 21 Mar 1911, *K. Dinter 2078* (syntype: BR! [fragments]); Büllsport, *K. Dinter s.n.* (syntype: not seen).

Cucumis angolensis Hooker filius ex Cogniaux var. *minor* Cogniaux & Harms, *Pflanzenr.* 88: 132. 1924.—TYPE: NAMIBIA. Okanbahe, Oct 1888, *G. Gürich 53* (holotype: B [presumed destroyed]).

Plants: Herbs or subshrubs (decumbent to suberect); perennial; with a woody rootstock; lacking tubers; monoecious. Stems procumbent or climbing (weakly); sulcate; not aculeate; shortly antrorse-strigose and pilose (both types of pubescence intermixed, or rarely not pilose); with nonbreakaway hairs. Stem hairs 0.2-0.7 or 1.2-1.6(-2.4) mm long. Nodes not geniculate. Internodes (1-)4-11 cm long. *Leaves*: Petioles 4-8.5 cm long (longer petioles found on herbaceous plants); sulcate; not aculeate; pubescence 2 different types uniformly intermixed on each petiole; shortly antrorse-strigose and pilose (with the basal hair cells enlarged); with nonbreakaway hairs. Leaf blades entire, pentalobate (weakly), or trilobate (only hastate leaves); with the margin serrate; very broadly ovate in outline; cordate (deeply so), or hastate at the base; with a basal sinus; (0.5-)1.2-2 cm deep; broadly acute or obtuse at the apex; (1.3-)4.5-10.5 X (1.5-)5-11 cm; 0.9 times longer than wide; hispidulous on the upper surface; not aculeate on the veins below; hispid on the veins below; hispidulous or hispid (rarely) on the intervenium below; with nonbreakaway hairs. Central leaf-blade lobe entire; shallowly ovate in outline; broadly acute or obtuse at the apex; (0.6-)1.5-2 X (0.9-)2.5-4.5 cm. Lateral leaf-blade lobes asymmetrical; entire; shallowly ovate or narrowly elliptic (only on hastate leaves) in outline; broadly acute at the apex; (0.3-)0.5-1(-2) X (0.4-)1.5-2.5 cm (longest and narrowest lateral lobes on hastate leaves). *Tendrils*: Present; solitary; simple; 2-7 cm long; not aculeate; antrorse-strigose. *Inflorescences*: Unisexual. *Male inflorescences*: Racemose, fasciculate, or a solitary flower; 1-11-flowered; sessile. *Male flowers*: Pedicel terete in cross section; 6.4-8 mm long; shortly antrorse-strigose and pilose (both types of pubescence intermixed); without bracteoles. Hypanthium infundibular; 4-4.8 mm long; 3.2-4 mm in diam; hispidulous or antrorse-strigose. Calyx lobes linear or narrowly triangular in outline; narrowly acute at the apex; 1.2-2.8 X 0.2-0.6 mm; hispidulous or antrorse-strigose. Corolla infundibular; hispidulous or puberulent outside; puberulent inside. Corolla tube 1.6-2.4 mm long; 4-5.6 mm in diam; hispidulous or puberulent outside. Corolla lobes elliptic in outline; mucronate and obtuse at the apex; 4-7.2 X 2.6-5.6 mm; hispidulous outside; puberulent inside. Stamens separating from the hypanthium 1.6-2 mm from the base of the hypanthium. Filaments terete in cross section; 0.4-0.8 mm long; 0.2-0.4 mm wide; glabrous. Anther thecae 2.4-2.6 mm long; puberulent. Anther connective narrowly oblong or obovate; unilobate or bilobate; obtuse at apex; papillate at the apex; 1-1.6 X 0.6-1.2 mm. Disc cylindrical; ca 1.2 mm long; 1.6-1.9 mm in diam. *Female inflorescences*: A solitary flower. *Female flowers*: Pedicel terete in outline; 3.2-4.8 mm long; shortly antrorse-strigose and pilose (both types of pubescence intermixed); with nonbreakaway hairs; cylindrical. The lower 2/3 of hypanthium fused to the ovary. Fused portion of hypanthium ellipsoid; 5.6-6.4 mm long; ca 4 mm in diam; not aculeate; lanate, antrorse sericeous, or puberulent (densely, hairs 2-2.8, 0.6-1.2, or ca 0.1 mm long, respectively). The upper 1/3 of hypanthium free from the ovary. Free portion of hypanthium 3.2-4 mm long; ca 2.8 mm in diam; not aculeate outside; hispid outside; puberulent inside. Calyx lobes linear in outline; narrowly acute at the apex; 1.2-1.4(-3.2) X ca 0.2 mm; hispid or hispidulous. Corolla hispid or hispidulous outside; puberulent inside. Corolla tube present; 2.8-4 mm long; ca 4.8 mm in diam; hispidulous outside. Corolla lobes elliptic to broadly elliptic in outline; mucronate and obtuse at the apex; 3.6-8 X 3.2-5.6 mm; hispidulous or hispid outside; puberulent inside. Staminodes present; separating from the free portion of the hypanthium 1.6-2 mm above the ovary; 0.3-0.4 mm long; ca 0.2 mm in diam; glabrous. Style ca 1.6 mm long; 0.3-0.6 mm in diam; subtended by a circular disc. Disc 0.3-0.6 mm long; 1.2-1.9 mm in diam. Stigma 2.2-2.4 mm long; ca 2 mm in diam; lobate; with 5 finger-like projections on the margin. Finger-like stigmatic projections 0.8-1.2 mm long; ca 0.6 mm in diam; papillate. *Fruit*: Not geocarpic, maturing above ground and readily visible. Pedicel terete in cross section; 1-1.5 cm long; shortly antrorse-strigose and pilose (both types of pubescence intermixed, many hairs broken off); with nonbreakaway hairs; cylindrical. Fruit in southern part of its range monocolored or in northern part of its range bicolored with longitudinal stripes from base to apex; green or yellow (pale, when ripe); with light green longitudinal stripes; globose; 2-2.5 cm long; 2-2.5 cm in diam; not aculeate; pilose and hispidulous (both types of pubescence intermixed), or hirsute; blunt at the apex. *Seeds*: Elliptic; 4.3-5 X 2.1-2.4 mm; 0.9-1.1 mm thick; unwinged. *Chromosome number*: 2n = 24 (Shimotsuma, 1965; Deakin, Bohn & Whitaker, 1971; Dane & Tsuchiya, 1976; Ramachandran, 1984; Yadava, Singh & Arya, 1984; Ramachandran & Narayan, 1985).

   Distribution: Africa: Angola (southwestern corner next to Namibia), Namibia, and South Africa (only northwestern Cape Province).

   Cucumis sagittatus ranges from the northwestern corner of Cape Province, South Africa, northwards through Namibia to the southwestern corner of Angola. Its distribution corresponds to the northern, coastal area of the Karoo-Namib Region as defined by White (1983). It is found in varying ecological conditions from extreme xeric to dry forest habitats. Four lines of variation are evident in this species: 1) Decumbent herbs versus suberect subshrubs; 2) leaves larger versus smaller; 3) leaf texture softer versus stiffer and more leathery; and 4) leaf bases deeply cordate versus hastate. From herbarium label data it appears that decumbent herbs with larger, softer leaves are encountered in more moderate environmental conditions. Hastate leaf bases are found on plants from the full spectrum of ecological conditions for the species. There is no correlation between either the characters themselves or between them and geographical distribution. Meeuse (1962) distinguished between monochrome fruit in the southern part of its distribution and bicolored fruit in the north. Fruit coloration may be worthy of formal taxonomic recognition at infraspecific rank. Further studies are needed to determine whether these are two stages in fruit development, immature fruit monochrome and mature fruit bicolored, or two mature fruit forms.

There are two identically labeled collections of *Wawra 263* in W. Since neither was indicated in anyway as *type* by Peyritsch, they are syntypes, and one is here designated as lectotype. I have considered the specimen of *Dinter 1436* in Cogniaux's personal herbarium, now deposited in BR and labelled in his handwriting, as the holotype of *C. dinteri*.

Cucumis subgen. *Cucumis*

Cucumis subgen. **Cucumis** Species 31 and 32.

Cucumis subgen. *Cucumis* has traditionally been called the 'Asian' element of *Cucumis*. *Cucumis* subgen. *Cucumis* is separated from the rest of the genus by its unique base chromosome number of $x = 7$; as would be expected it has not been successfully crossed with any member of subgen. *Melo*. It is also distinguished by a unique flavonoid compound (Brown et al., 1969) and the presence of Cucurbitacin C (Enslin and Rehm, 1957; Rehm, 1960). Vegetatively, members of this subgenus have a characteristic appearance that is difficult to describe. The only distinguishing characteristic that can be described is the shape of the leaf-blade lobes, some form of triangular in *Cucumis* subgen. *Cucumis* versus various forms of elliptic, oblong, and ovate in *Cucumis* subgen. *Melo*. All the biosystematic data is based on *C. sativus*, and nothing is known about *C. hystrix* other than morphology. *Cucumis hystrix* is associated with *C. sativus* because of morphological similarity. Biosystematic data are urgently needed for *C. hystrix* to adequately assess the situation.

31. **Cucumis sativus** Linnaeus, Sp. pl. *ed. 1*, 1012. 1753.--TYPE: EUROPE. Cultivated plants, *J. Burser vol. 17, no. 97* (lectotype: UPS [Pas et al., 1985]).

All synonyms are listed in Appendix 2.

Plants: Herbs; annual; without a woody rootstock; lacking tubers; monoecious. Stems procumbent or climbing; sulcate; not aculeate; hispid to retrorse-strigose (strongly so in Thailand); with nonbreakaway hairs. Stem hairs 0.4-2 mm long. Nodes not geniculate. Internodes 4-12 cm long. *Leaves*: Petioles 2.5-7.5(-22) cm long; sulcate; not aculeate; pubescence a single type on each petiole; hispid to retrorse-strigose; with nonbreakaway hairs. Leaf blades pentalobate or heptalobate; with the margin serrate; broadly ovate in outline; cordate at the base; with a basal sinus; 1.5-4 cm deep; narrowly acute to acute at the apex; 6-15(-40) X 6-15(-35) cm; 1-1.2 times longer than wide; hispid on the upper surface; not aculeate on the veins below; hispid on the veins below; hispid on the intervenium below; with nonbreakaway hairs. Central leaf-blade lobe entire; broadly triangular to very broadly triangular to shallowly triangular in outline; narrowly acute to acute at the apex; 4-6.5(-16) X 4.5-8.5(-21) cm. Lateral leaf-blade lobes symmetrical or asymmetrical; entire; shallowly triangular in outline; broadly acute at the apex; 0.5-3(-6) X 2-6(-16) cm. *Tendrils*: Present; solitary; simple; 7-12(-30) cm long; not aculeate; hispidulous or antrorse-strigose. *Inflorescences*: Unisexual. *Male inflorescences*: Fasciculate or a solitary flower (rarely); 1- or 3-7(-10)-flowered; sessile. *Male flowers*: Pedicel sulcate or terete in cross section; 5-15(-18) mm long; hispidulous or hispid; without bracteoles. Hypanthium campanulate; 5.4-6.3 mm long; 3.6-4.5 mm in diam; pilose.

Calyx lobes narrowly oblong to linear in outline; narrowly acute at the apex; 2.7-4.1(-5.1) X ca 0.4 mm; sparsely pilose. Corolla infundibular; pilose and puberulent outside (pilose on the central veins of the corolla lobes and puberulent on the lateral veins); glabrous inside. Corolla tube 3.4-4.9 mm long; 7-9 mm in diam; pilose and puberulent outside (pilose on the central veins of the corolla lobes and puberulent on the lateral veins). Corolla lobes elliptic in outline; obtuse and mucronate at the apex; 7-18(-22) X 5-11(-13) mm; pilose and puberulent outside (pilose on the central vein and puberulent on the lateral veins); glabrous inside. Stamens separating from the hypanthium 2-2.5 mm from the base of the hypanthium. Filaments terete in cross section; 0.4-0.8 mm long; 0.3-0.5 mm wide; glabrous. Anther thecae 2.5-2.8 mm long; puberulent. Anther connective obovate; unilobate or bilobate; obtuse at apex; papillate at the apex; 1.6-2 X 0.6-1 mm. Disc cylindrical; 1-1.4 mm long; 1-1.5 mm in diam. *Female inflorescences*: A solitary flower or fascicles (rarely, on cultivated specimens); 1(-3)-flowered. *Female flowers*: Pedicel terete in outline; 2-18(-20) mm long; hispidulous or hispid; with nonbreakaway hairs; cylindrical. The lower 3/4 of hypanthium fused to the ovary. Fused portion of hypanthium ellipsoid or ovoid; 9.5-10.5 mm long; ca 3.5 mm in diam; not aculeate or aculeate; glabrous or hirsute. Aculei soft (when present); 1.4-1.9 mm long. Hyaline bristle at apex of aculeus 1-1.4 mm long; 1.4-2.8 times as long as the opaque base. The upper 1/4 of hypanthium free from the ovary. Free portion of hypanthium 3.2-3.6 mm long; 2.8-3.2 mm in diam; not aculeate outside; pilose outside; glabrous or puberulent (minutely) inside. Calyx lobes narrowly elliptic in outline; narrowly acute at the apex; 2.8-3.6(-12) X 0.4-0.5(-3) mm; sparsely pilose. Corolla pilose and puberulent outside (pilose on the central veins of the corolla lobes and puberulent on the lateral veins); glabrous inside. Corolla tube present; 3.5-6.5 mm long; 5.5-7 mm in diam; pilose and puberulent outside (pilose on the central veins of the corolla lobes and puberulent on the lateral veins). Corolla lobes broadly elliptic in outline; obtuse and mucronate at the apex; 10-15(-25) X 9.5-15(-18) mm; pilose and puberulent outside (pilose on the central veins and puberulent on the lateral veins); glabrous inside. Staminodes present; separating from the free portion of the hypanthium 1.4-1.6 mm above the ovary; 0.5-0.6 mm long; 0.2-0.4 mm in diam; glabrous. Style ca 1.6 mm long; ca 0.6 mm in diam; subtended by a circular disc. Disc 0.6-0.8 mm long; 1.4-2.4 mm in diam. Stigma 2.8-3.2 mm long; ca 2.4 mm in diam; lobate; with 3 finger-like projections on the margin. Finger-like stigmatic projections 1.6-2 mm long; 0.8-1.2 mm in diam; papillate. *Fruit*: Not geocarpic, maturing above ground and readily visible. Pedicel sulcate in cross section; 1.5-3 cm long; hispidulous or hispid; with nonbreakaway hairs; cylindrical. Fruit monocolored or bicolored with longitudinal stripes from base to apex (with the surface smooth or netted); green (immature or mature), white (immature or mature), yellow (mature only), or brown (mature only); with light green or white longitudinal stripes; ellipsoid to cylindrical; 5-20(-50) cm long; 3-5(-12) cm in diam; not aculeate; glabrous; blunt at the apex. *Seeds*: Elliptic to ovate; 7.2-11.2 X 3.2-4.9 mm; 1.3-1.6 mm thick; unwinged. *Chromosome number*: 2n = 14 (Bhaduri & Bose, 1947; Naithani, 1947; Shimotsuma, 1965; Deakin, Bohn & Whitaker, 1971; Ramachandran, 1984; Yadava, Singh & Arya, 1984; Ramachandran & Narayan, 1985).

   Distribution: Asia: Burma, China (Yunnan Province), India, Sri Lanka, and Thailand. Introduced: cultivated worldwide.

   Jeffrey (1980) was first to lectotypify *C. sativus* by the phrase name in *Hortus cliffortianus* (Linnaeus, 1738). His selection was sanctioned by the *International Code of Botanical Nomenclature* (ICBN) (Stafleu et al., 1978) in effect at the time. Pas et al. (1985) rejected Jeffrey's lectotypification on the grounds that "the use of a phrase name as a lectotype is undesirable as it fails to establish the application of the name adequately," and they selected *J. Burser vol. 17, no. 97*. The ICBN in effect at that time (Voss et al., 1983), stipulated that a lectotype could be superseded only for the following reasons: 1) If the holotype or any of the original material was rediscovered; 2) if the lectotype conflicted with the protologue; 3) if the lectotype was selected by a mechanical method; or 4) if the lectotype was a mixed collection. Obviously, Jeffrey's lectotype did not suffer from any of these qualifications, so Pas et al. had no grounds under the ICBN then in effect to reject it. In the next ICBN (Greuter et al., 1988), which became effective in 1987, the definition of lectotype was altered to exclude descriptions as lectotypes. Consequently Jeffrey's lectotype of *C. sativus* was invalidated in 1987 and that of Pas et al. became valid at that time.

Variation within *C. sativus* is extensive. It is not as great as in *C. melo*, but it is wider than in any non-domesticated species of *Cucumis*. The fused portion of the hypanthium on female flowers of specimens from Xishuangbannen autonomous region of Yunnan Province, China, designated as *C. sativus* var. *xishuangbannanesis*, ranges from aculeate to hirsute or very rarely glabrous. These glabrous specimens from China are also characterized by orange-fleshed fruits. Qi et al. (1983) have shown this unusual fruit coloration to be controlled by a single dominant gene. Cultivars have a broad range of sexual expression: Monoecious, gynoecious, andromonoecious, androdiecious, hermaphroditic, and others (J.E. Staub, pers. comm.). Only monoecism could be identified with confidence in the dried specimens, so it alone is cited in the description of the species. The aculei on the female hypanthium vary in the development of their opaque bases, which range from 0.4 to almost 1 mm long. When the base does not develop at all, the specimen appears hirsute because the hyaline apical bristle is persistent. *Cucumis setosus* lacks the opaque base of the aculei so that it appears to be hirsute. This character is found frequently among cultivated specimens because agronomists have selected for the lack of aculei and smooth fruits.

Smaller, more delicate plants with bitter fruits have traditionally been identified as *C. sativus* var. *hardwickii*. The typical members of this variety are easily identified, but none of the morphological characters now available clearly separate it from the rest of the species. The size character would have to be established arbitrarily because there are many specimens intermediate in size between typical var. *hardwickii* and the cultivated forms. *Cucumis sativus* var. *hardwickii* can be distinguished from the remainder of the species by isozyme patterns (Knerr and Staub, 1991), and morphological types have been recognized in var. *harwickii* (Schuman et al., 1985; Staub and Kupper, 1986).

32. **Cucumis hystrix** Chakravarty, J. Bombay Nat. Hist. Soc. 50(4): 896. 1952.—TYPES: INDIA. MEGHALAYA: Garo Hills, Tura Mountain, 4 Nov 1929, *N.E. Parry 859* (holotype: K).—ARUNACHAL PRADESH: East Bengal, Mishmee, *Griffith 2554* (paratype: K).

   Cucumis muriculatus Chakravarty, J. Bombay Nat. Hist. Soc. 50(4): 896. 1952.—TYPE: BURMA. Ruby Mines District, Oct 1912, *J.H. Lace 6325* (holotype: E; isotype: E).

   Cucumis yunnanensis C.Y. Wu in C. Jeffrey, The Cucurbitaceae of Eastern Asia 22. 1980. Invalid, Jeffrey cited it as a synonym of *C. hystrix* Chakravarty.

Plants: Herbs; monoecious. Stems procumbent; sulcate; not aculeate; antrorse-strigose; with nonbreakaway hairs. Stem hairs 0.3-0.8(-1.2) mm long. Nodes not geniculate. Internodes 6-8(-12) cm long. *Leaves*: Petioles 2.5-5(-12) cm long; sulcate; not aculeate; pubescence a single type on each petiole; retrorse-strigose; with nonbreakaway hairs. Leaf blades pentalobate; with the margin serrate; ovate to broadly ovate in outline; cordate at the base; with a basal sinus; 1.5-2.5(-4) cm deep; narrowly acute at the apex; 7.5-13(-15) X 5.5-12(-14) cm; 1.1-1.4 times longer than wide; hispidulous to antrorse-strigose on the upper surface; not aculeate on the veins below; antrorse-strigose to hispidulous on the veins below; hispidulous to antrorse-strigose on the intervenium below; with nonbreakaway hairs. Central leaf-blade lobe entire; broadly triangular to very broadly triangular in outline; narrowly acute at the apex; 3.5-8 X 3.5-9(-11) cm. Lateral leaf-blade lobes symmetrical or asymmetrical; entire; shallowly triangular or broadly triangular in outline; narrowly acute, broadly acute or obtuse at the apex; 0.7-3 X 2-2.8 cm. *Tendrils*: Present; solitary; simple; 3-5(-11) cm long; not aculeate; antrorse-strigose. *Inflorescences*: Unisexual. *Male inflorescences*: Fasciculate or racemose (shortly so); 5-7-flowered; sessile. *Male flowers*: Pedicel terete in cross section; 1.6-5.6 mm long; antrorse-strigose; without bracteoles. Hypanthium infundibular or campanulate; 2.4-5.6 mm long; 1.6-2.8 mm in diam; hispidulous or antrorse-strigose. Calyx lobes narrowly oblong in outline; narrowly acute at the apex; 1-2 X 0.2-0.4 mm; antrorse-strigose. Corolla infundibular; antrorse-strigose (on the veins) outside; glabrous inside. Corolla tube 1.6-1.8 mm long; ca 2.5 mm in diam; antrorse-strigose (on the veins) outside. Corolla lobes elliptic to broadly elliptic in outline; obtuse and mucronate at the apex; ca 4 X 3.2-3.6 mm; antrorse-strigose (on the veins) outside; glabrous inside. Stamens separating from the hypanthium ca 2.8 mm from the base of the hypanthium. Filaments terete in cross section; ca 1 mm long; ca 0.4 mm wide; glabrous. Anther thecae 2.1-2.2 mm long; puberulent. Anther connective obovate; unilobate; obtuse at apex; papillate at the apex; 1.4-1.6 X ca 0.6 mm. Disc cylindrical; ca 0.9 mm long; ca 1.1 mm in diam. *Female inflorescences*: A solitary flower. *Female flowers*: Pedicel terete in outline; 1.2-4 mm long; antrorse-strigose; with nonbreakaway hairs; cylindrical. The lower 2/3 of hypanthium fused to the ovary. Fused portion of hypanthium ellipsoid to ovoid; 8-12 mm long; 2.4-3.2 mm in diam; aculeate; glabrous. Aculei soft to stiff; 1.4-2.6 mm long. Hyaline bristle at apex of aculeus 1-2 mm long (with the base enlarged); 2.5-3.3 times as long as the opaque base. The upper 1/3 of hypanthium free from the ovary. Free portion of hypanthium 4-6.4 mm long; 2.4-3.2 mm in diam; not aculeate outside; hispidulous to antrorse-strigose outside; puberulent inside. Calyx lobes narrowly oblong in outline; narrowly acute at the apex; 1.2-3.2 X 0.3-0.5 mm; antrorse-strigose. Corolla antrorse-strigose (on the veins) outside; glabrous or puberulent inside. Corolla tube present; 1.6-2.4 mm long; ca 2.4 mm in diam; antrorse-strigose (on the veins) outside. Corolla lobes broadly elliptic in outline; obtuse and mucronate at the apex; 4.4-4.9 X 4-4.4 mm; antrorse-strigose (on the veins) outside; short puberulent inside. Staminodes present; separating from the free portion of the hypanthium 2-2.4 mm above the ovary; 0.8-1 mm long; ca 0.2 mm in diam; glabrous. Style 2-2.4 mm long; 0.2-0.6 mm in diam; subtended by a circular disc. Disc ca 1 mm long; ca 2.8 mm in diam. Stigma 1.6-2.4 mm long; 1.2-1.6 mm in diam; lobate; with 3 finger-like projections on the margin. Finger-like stigmatic projections ca 1.6 mm long; ca 0.8 mm in diam; papillate. *Fruit*: Not geocarpic, maturing above ground and readily visible. Pedicel terete in cross section; 0.3-0.8 cm long; antrorse-strigose; with nonbreakaway hairs; cylindrical. Fruit green (when immature) or yellow (when mature); ellipsoid to obovoid; 2-5 cm long; 1-2 cm in diam; aculeate; glabrous; rostrate at the apex. Beak 0.2-0.5 cm long. Aculei terete; with hyaline bristle lost (the enlarged base of the bristle persistent); 0.7-2.1 mm long; 0.2-0.4 mm in diam. *Seeds*: Seeds ovate; 3.3-5 X 1.9-2.6 mm; 0.6-0.7 mm thick; unwinged.

Distribution: Asia: Burma, China (Yunnan), India (Assam), and Thailand.

Cucumis hystrix resembles *C. sativus* in its vegetative and floral morphology, but is easily distinguished by its mature fruit. The short aculei on the fused portion of the female hypanthium continue to grow as the fruit develops, so the mature fruit are covered by prominent aculei.

In India, *C. hystrix* has only been collected northeast of Bangladesh in the Brahmaputra River valley. Most of the collections were made in the nineteenth century with only two from this century. Only a single collection is known from Burma, the holotype gathering collected in 1912. The Chinese collections are all from the Province of Yunnan, and were made from 1900 to 1939. The Thai collections are from widely dispersed areas within the country. The earliest one was made in 1914, and the last available ones are from 1958.

Doubtful Names

Cucumis angulatus Forsskål, Fl. aegypt.-arab. 168. 1775.
Cucumis anheimta Forsskål, nom. nud., Fl. aegypt.-arab. cxxii. 1775.
Cucumis arenarius Schrader, Linnaea 12: 416. 1838. Illegitimate, later homonym of *Cucumis arenarius* Schumacher & Thonning. Schrader (1838) referred to his species as, "Vix variet. local. C. africani."
Cucumis barbatus A.W. Roth, Catal. bot. 2: 122. 1800.
Cucumis daucus-indicus Forsskål, nom. nud., Fl. aegypt.-arab. cxxii. 1775.
Cucumis dissectifolius Naudin var. *filiformis* Cogniaux in Schinz, Verh. Bot. Vereins Prov. Brandenburg 30: 150.   1889.—TYPE: NAMIBIA. Olukonda, intribu Ondonga (Amboland), *H. Schinz s.n.* (holotype).
Cucumis echinaceus Sprenger, Bull. Soc. Tosc. Ortic. 19: 115. 1894.
Cucumis farinosa Ehrenberg ex Dragendorff, nom. nud., Die Heilpflanzen der Verschiedenen Völker und Zeiten 651. 1898.
Cucumis gurmia Wallich, nom. nud., Numer. List, number 6726. 1828.
Cucumis hilapikku Wallich, nom. nud., Numer. List, number 6754bis. 1828.
Cucumis inedulis Forsskål, nom. nud., Fl. aegypt.-arab. cxxii. 1775.
Cucumis japonicus Siebold, nom. nud., Synopsis plantarum oeconomicarum universi regni japonici, as an entry in the Itabula synoptica usus plantarum. 1830.
Cucumis melopepo Steudel, nom. nud., Nomencl. bot. ed. 2, 2: 303. 1840.
Cucumis missionis Wallich, nom. nud., Numer. List, number 6728. 1828.
Cucumis ollahie Forsskål, nom. nud., Fl. aegypt.-arab. cxxii. 1775.
Cucumis melopepo Steudel, nom. nud., Nomencl. bot. ed. 2, 2: 303. 1840.
Cucumis missionis Wallich, nom. nud., Numer. List, number 6728. 1828.
Cucumis ollahie Forsskål; nom. nud., Fl. aegypt.-arab. cxxii. 1775.
Cucumis orientalis Forsskål, nom. nud., Fl. aegypt.-arab. cxxii. 1775.
Cucumis perpusillus Noroña, nom. nud., Verh. Batav. Genootsch. Kunst. Wet. 5(4): 11. 1790.
Cucumis purpureus Noroña, nom. nud., Verh. Batav. Genootsch. Kunst. Wet. 5(4): 12. 1790.
Cucumis tuberculatus Forsskål, nom. nud., Fl. aegypt.-arab. cxxii. 1775.

Excluded Names

Cucumis acutangulus Linnaeus, Sp. pl. ed. 1, 1011.   1753.—TYPE: (neotype: LINN, sheet number 1152.7!).   = *Luffa acutangula* (Linnaeus) Roxburgh (Jeffrey, 1980).
Cucumis amarissimus Schrader, Linnaea 12: 414. 1838. Invalid, cited as synonym of *Citrullus amarus* Schrader.   = *Citrullus lanatus* (Thunberg) Matsumura & Nakai (Fursa and Filov, 1982).
Cucumis anguinus Linnaeus, Syst. nat. ed. 10 1279.   1759.   = *Trichosanthes cucumerina* Linnaeus (Chakravarty, 1959; Jeffrey, 1980).

Cucumis argyi H. Léveillé, Mem. Real Acad. Ci. Barcelona 12(22): 8. 1916.—TYPE: CHINA. Jiangsu: *P.C. d'Argy s.n.* (holotype: E). = *Momordica charantia* Linnaeus (Jeffrey, 1980; Lu and Zhang, 1986).

Cucumis asininus Garsault, Fig. pl. méd. *t. 237.* 1764; Expl. abr. *pl. 154.* 1765.—TYPE: *t. 237,* Fig. pl. méd., 1764 (holotype). = *Ecbalium elaterium* (Linnaeus) A. Richard (Cogniaux and Harms, 1924).

Cucumis asper Cogniaux, Bull. Herb. Boissier, sér. 2, 1: 882. 1901.—TYPES: NAMIBIA. Hereroland, Abasis, *K. Dinter 1447* (holotype: BR!; isotype: Z). = *Cucumella aspera* (Cogniaux) C. Jeffrey (Jeffrey, 1962).

Cucumis bipinnatifidus R. Wight ex Naudin, Ann. Sci. Nat. Bot., sér. 4, 12: 99. 1859.—TYPE: India, *R. Wight s.n.* (holotype: K). Invalid, cited as synonym of *Citrullus colocynthis* (Linnaeus) Schrader.

Cucumis bicirrha Forster ex Guillemin, Ann. Sci. Nat. Bot., sér. 2, 7: 350. 1837. = *Lagenaria siceraria* (Molina) Standley (Cogniaux and Harms, 1924).

Cucumis cecili N.E. Brown, Bull. Misc. Inform. 1906: 104. 1906.—TYPE: ZIMBABWE. Manika District: Inyanga Mountains, 1800-2100 m., *E. Cecil 225* (holotype: K). = *Oreosyce africana* Hooker filius (Jeffrey, 1967).

Cucumis chrysocomus Schumacher, Beskr. Guin. pl. 201. 1827.—TYPE: GHANA. *P. Thonning s.n.* (holotype: C). = *Raphidiocystis chrysocoma* (Schumacher) C. Jeffrey (Jeffrey, Mar 1962, 1967; Meeuse, May 1962; Keraudren-Aymonin, 1975).

Cucumis chrysocarpa Schumacher ex Welwitsch, nom. nud., Apont. 589. 1859. This epithet is the result of a typographical error transliterating *chrysocomus* into *chrysocarpa*.

Cucumis citrullus Dumortier, nom. nud., *Fl. belg.* 54. 1827. = *Citrullus* sp.

Cucumis citrullus (Linnaeus) Seringe in de Candolle, Prodr. 3: 301. 1828.—BASIONYM: *Cucurbita citrullus* Linnaeus. = *Citrullus lanatus* (Thunberg) Matsumura & Nakai (Jeffrey, 1967).

Cucumis citrullus (Linnaeus) Seringe var. *pasteca* Seringe in de Candolle, Prodr. 3: 301. 1828. = *Citrullus lanatus* (Thunberg) Matsumura & Nakai.

Cucumis citrullus (Linnaeus) Seringe var. *jace* Seringe in de Candolle, Prodr. 3: 301. 1828. = *Citrullus lanatus* (Thunberg) Matsumura & Nakai.

Cucumis colocynthis Linnaeus, Sp. pl. ed. 1, 1011. 1753.—TYPE: SWEDEN. Plant cultivated at Uppsala, Sweden (lectotype: LINN, sheet number 1152.1!). = *Citrullus colocynthis* (Linnaeus) Schrader (Jeffrey, 1967, 1980).

Cucumis cordifolius E. Meyer ex Drège, nom. nud., Zwei pflanzgeogr. Dokum. 149, 176. 1843; Sonder in Harvey & Sonder, as synonym of *Momordica cordifolia* E. Meyer ex Sonder, Fl. cap. 2: 492. 1862. = *Momordica foetida* Schumacher (Meeuse, 1962).

Cucumis courtoisii H. Léveillé, Mem. Real Acad. Ci. Barcelona 12(22): 8. 1916.—TYPE: CHINA. Jiangsu: Vou Sié, *P.C. d'Argy s.n.* (holotype: E). = *Thladiantha nudiflora* Hemsley ex Forbes & Hemsley (Jeffrey, 1980; Lu and Zhang, 1986).

Cucumis crassiflorus Sessé & Mociño, Fl. mexic. 249. 1896. Its tendrils are described as 3-, 4-, or 5-fid. = *Sicana odorifera* (Velloso) Naudin (Jeffrey, pers. comm.).

Cucumis dissectus Decaisne, Nouv. Ann. Mus. Hist. Nat. 3: 449. 1834. = *Citrullus lanatus* (Thunberg) Matsumura & Nakai (Fursa and Filov, 1982).

Cucumis edulis Spach ex Steudel, nom. nud., Nomencl. bot. ed. 2, 1: 451. 1840. = *Citrullus lanatus* (Thunberg) Matsumura & Nakai (Fursa and Filov, 1982).

Cucumis foetidissima Humboldt, Bonpland & Kunth ex Hemsley, Biol. cent.-amer., Bot. 1: 483. 1880. Invalid, cited as synonym of *Cucurbita perennis* A. Gray.

Cucumis fricatorius Sessé & Mociño, Fl. mexic. 249. 1896. Its tendrils are described as 4-fid. = *Luffa operculata* (Linnaeus) Cogniaux in C. Martius (Jeffrey, pers. comm.).

Cucumis glaber T. Walter, Fl. carol. 240. 1788. = *Melothria pendula* Linnaeus (Cogniaux and Harms, 1924).

Cucumis himalensis Royle, nom. nud., Ill. bot. Himal. Mts. 1: 218. 1835.—TYPES: INDIA. Simla, *Royle 80/48* (syntype: LIV!); near Broang Pass, *Maxwell s.n.* (syntype). *Royle 80/48* has bifid tendrils and racemose male inflorescences with the peduncle ca 17 cm long.

Cucumis integrifolius Roxburgh, Fl. Ind. 3: 724. 1832.—TYPE: INDIA. *Ic. Roxb.* (neotype: K [proposed in Jeffrey (1980)]). = *Gymnopetalum integrifolium* (Roxburgh) Kurz (Jeffrey, 1980).

Cucumis intermedius M. Roemer, Fam. nat. syn. monogr. 2: 80. 1846.—TYPE: Bot. Reg. t. 980. 1826 (holotype). = *Momordica charantia* Linnaeus (Cogniaux and Harms, 1924; Chakravarty, 1959).

Cucumis laciniosa Schrader, Linnaea 12: 414. 1838. Invalid, cited as synonym of *Citrullus amarus* Schrader. = *Citrullus lanatus* (Thunberg) Matsumura & Nakai (Fursa and Filov, 1982).

Cucumis lagenarius Dumortier, nom. nud., Fl. belg. 54. 1827. = *Lagenaria siceraria* (Molina) Standley.

Cucumis lineatus Bosc, J. Hist. Nat. 2: 251, *t. 37.* 1792.—TYPE: FRENCH GUIANA. Cayenne, *Nectou s.n.* (holotype: P). Its tendrils are described and illustrated as 3-fid, and it is described as next to *Cucumis acutangulus* Linnaeus (= *Luffa acutangula* (Linnaeus) Roxburgh) in the system of classification.

Cucumis maderaspatanus Linnaeus, Sp. pl. ed. 1, 1012. 1753.—TYPE: *t. 170, f. 2, Cucumis maderaspatensis fructu minimo* Plukenet, Phytographia 3. 1691. (lectotype [Meeuse, 1962]). = *Mukia maderaspatanus* (Linnaeus) M. Roemer (Jeffrey, 1980).

Cucumis mairei H. Léveillé, Cat. pl. Yun-Nan. 64. 1916.—TYPE: CHINA. Yunnan: Dongchuan, 2500 m., Sep 1912, *E.E. Maire s.n.* (holotype: E). = *Lagenaria siceraria* (Molina) Standley (Jeffrey, 1980; Lu and Zhang, 1986).

Cucumis megacarpus G. Don, Gen. hist. 3: 28. 1834. = *Luffa* sp.

Cucumis membranifolius Hooker filius in Oliver, Fl trop. Afr. 2: 545. 1871.—TYPE: ETHIOPIA. Tigre, *G.H.W. Schimper 1457* (holotype: K) = *Oreosyce africana* Hooker filius (Jeffrey, 1967).

Cucumis muelleri Naudin, Ann. Sci. Nat. Bot., sér. 4, 11: 84. 1859. Illegitimate, it was nomenclaturally superfluous because it included the type of a name, *Cucurbita micrantha* F. Mueller, which ought to have been adopted. = *Mukia micrantha* (F. Mueller) F. Mueller (Telford, 1982).

Cucumis naudinianus Sonder in Harvey & Sonder, Fl. cap. 2: 496. 1862.—TYPES: SOUTH AFRICA. Transvaal: Potchefstroom, Mooi River, *J. Burke 488* (syntype: K); Magalisberg, *C. Zeyher 586* (syntype: K). = *Acanthosicyos naudinianus* (Sonder) C. Jeffrey (Jeffrey, 1975).

Cucumis odoratissimus Sessé & Mociño, Fl. mexic. 249. 186. Its tendrils are described as 3-, 4-, or 5-fid. = *Sicana odorifera* (Velloso) Naudin (Jeffrey, pers. comm.).

Cucumis parvifolius Cogniaux, J. Linn. Soc., Bot. 29(197): 19. 1891.—TYPE: MADAGASCAR. Amongst grasses near Lake Itasy (Interior), *G.F. Scott Elliot 1947* (holotype: BR). = *Oreosyce africana* Hooker filius (Keraudren, 1968).

Cucumis pavel Kosteletzky, Allg. med.-pharm. Fl. 2: 738. 1833.—TYPE: Rheede, Hort. malab. 8: 27-28, *t. 14.* 1688. = *Coccinia grandis* (Linnaeus) Voigt (Nicolson et al., 1988).

Cucumis pedatus Linnaeus, Syst. nat. ed. 10, 1279. 1759.--TYPE: Plumier, Pl. amer. 13, *t. 23.* 1755. = *Psiguria pedata* (Linnaeus) R.A. Howard (Howard, 1973).

Cucumis pentandrus Roxburg ex Wight & Arnott, Prodr. fl. Ind. orient. 1: 343. 1834.—TYPE: Icones Roxburghianae *t. 459* (K!). Invalid, cited as synonym of *Luffa pentandra* Roxburgh. = *Luffa cylindrica* (Linnaeus) M.J. Roemer (Jeffrey, pers. comm.).

Cucumis pepo Dumortier, nom. nud., Fl. belg. 54. 1827. = *Cucurbita* sp.

Cucumis perennis E. James, Account exped. Pittsburgh 2: 346. 1823. = *Cucurbita foetidissima* Kunth (Bailey, 1943).

Cucumis operculatus Roxburg ex Wight & Arnott, Prodr. fl. Ind. orient. 1: 343. 1834.—TYPE: Icones Roxburghianae *t. 460* (K!). Invalid, cited as synonym of *Luffa amara* Roxburgh. = *Luffa* sp.

Cucumis sepium G. Meyer, Prim. fl. esseq. 278. 1818. = *Luffa operculata* (Linnaeus) Cogniaux in C. Martius (Cogniaux & Harms, 1924; Heiser et al., 1988).

Cucumis striatus A. Richard, Tent. fl. abyss. 1: 295. 1847. = *Coccinia adoënsis* (A. Richard) Cogniaux in de Candolle (Cogniaux and Harms, 1924). P was examined in 1987, and the type of *C. striatus* could not be found.

Cucumis subsericeus Hooker filius in Oliver, Fl. trop. Afr. 2: 545. 1871.—TYPE: ANGOLA. Cuanza Norte: Pungo Andongo, *F.M.J. Welwitsch 838* (holotype: BM). = *Oreosyce africana* Hooker filius (Jeffrey, 1967; Halliday and Jeffrey, 1978).

Cucumis trifoliatus Linnaeus, Syst. nat. ed. 10, 1279. 1759. = ? *Psiguria* sp. (Cogniaux and Harms, 1924).

Cucumis trilobatus Linnaeus, Syst. nat. ed. 10, 1279. 1759.--TYPE: Plumier, Pl. amer. *t. 22*. 1755. = *Psiguria trilobata* (Linnaeus) R.A. Howard (Howard, 1973).

Cucumis tuberosus Roxburg ex Wight & Arnott, Prodr. fl. Ind. orient. 1: 344. 1834.—TYPE: Icones Roxburghianae *t. 461* (K!). Invalid, cited as synonym of *Luffa tuberosa* Roxburgh. = *Momordica cymbalaria* Hooker filius (Jeffrey, 1980).

Cucumis umbrosus Meeuse & Strey in Meeuse, Bothalia 8(1): 67. 1962.—TYPES: SOUTH AFRICA. Transvaal: Soutpansburg District, Williespoort, 8 Feb 1959, *A.D.J. Meeuse 10612* (paratypes: BR! LISC! PRE!); Middelburg District, wooded kloof S of Loskop Dam, ca 3m from Loskop Dam Rest Camp, 3 Oct 1958, *A.D.J. Meeuse & R.G. Strey 10317* (holotype: PRE!; isotypes: B BOL BM BR! EA K L M P S! SRGH); Middelburg District, near waterfall, on steep rock cliff opposite [Loskop Dam Rest] Camp, 19 Dec 1955, *R.G. Strey 2851* (paratypes: K! L! PRE! SRGH!). = *Cucumella bryoniifolia* (Merxmüller) C. Jeffrey (Halliday and Jeffrey, 1978).

Cucumis vulgaris (Schrader) E.H.L. Krause, J. Sturm's Deutschl. Fl. ed. 2, 12: 246. 1906.—BASIONYM: *Citrullus vulgaris* Schrader. = *Citrullus lanatus* (Thunberg) Matsumura & Nakai (Fursa and Filov, 1982).

LITERATURE CITED

Adanson, M. 1763. *Fam. pl.*, vol. 2. Paris: Vincent.

Alefeld, F.G.C. 1866. *Landw. Fl.* Berlin: Wiegandt & Hempel.

Andersen, J.S. 1977. Cucurbitaceae. In *Fl. Iranica* 123: 1-14. Graz: Akademische Druck- und Verlagsanstalt.

Andrus, C.F., and G. Fassuliotis. 1965. Crosses among *Cucumis* species. *Veg. Improv. Newslett.* 7: 3.

Anonymous. 1984. *Webster's New Geographical Dictionary.* Springfield: Merriam-Webster Inc.

Anonymous. 1985. *The Times Atlas of the World, ed. 7.* London: Times Books.

Anonymous. 1990. *Agricultural statistics, United States Department of Agriculture.* Washington: Government Printing Office.

Arnott, G.A.W. 1841. On the Cucurbitaceae. *J. Bot.* (Hooker) 3: 271-280.

Ayyangar, K.R. 1967. Taxonomy of Cucurbitaceae. *Bull. Natl. Inst. Sci. India* 34: 380-396.

Bailey, L.H. 1943. Article 6. Species of *Cucurbita. Gentes Herb.* 6(5): 267-322.

Batra, S. 1953. Interspecific hybridisation in the genus *Cucumis. Sci. & Cult.* 18(9): 445-446.

Bauhin, C. 1623. *Pinax.* Sumptibus & typis Ludovici Regis.

Bauhin, J. 1650. *Hist. pl.*, vol. 2. Yverdon.

Bhaduri, P.N., and P.C. Bose. 1947. Cyto-genetical investigations in some common cucurbits, with special reference to fragmentation of chromosomes as a physical basis of speciation. *J. Genet.* 48: 237-256.

Berhaut, J. 1954. *Fl. Sénégal.* Dakar: Éditions Clairafrique.

Berhaut, J. 1967. *Fl. Sénégal ed. 2.* Dakar: Éditions Clairafrique.

Berhaut, J. 1975. Cucurbitaceae. In *Fl. ill. Sénégal* 4: 226-309. Dakar: Gouvernement du Sénégal.

Brickell, C.D., et al. 1980. International code of nomenclature for cultivated plants—1980. *Regnum Veg.* 104: 1-32.

Britton, N.L., and P. Wilson. 1925. *Cucumis. Bot. Porto Rico* 6(2): 264. New York: N.Y. Academy of Sciences.

Brown, G.B., J.R. Deakin, and M.B. Wood. 1969. Identification of *Cucumis* species by paper chromatography of flavonoids. *J. Amer. Hort.* Soc. 94: 231-234.

Carreira, A. 1968. As Companhias Pombalinas de navegação, comércio e tráfico de escravos entre a costa africana e o nordeste brasileiro. *Bol. Cult. Guiné Portug.* 23(89-90): 5-88; 23(91-92): 89-565.

Chandola, R.P., M.P. Bhatnagar, and I. Tokuta. 1965. Cytological studies of some varieties of musk melon with special reference to their relationship. *Cytologia* 30(3): 252-259.

Chakravarty, H.L. 1946. Studies on Indian Cucurbitaceae with special remarks on distribution and uses of economic species. *Indian J. Agric. Sci.* 16(1): 1-90, *t. 1-12.*

Chakravarty, H.L. 1958. Morphology of the staminate flowers in the Cucurbitaceae with special reference to the evolution of the stamen. *Lloydia* 21(2): 49-87.

Chakravarty, H.L. 1959. Monograph of Indian Cucurbitaceae. *Rec. Bot. Surv. India* 17(1): 1-234.

Chakravarty, H.L. 1961. Geographical range of Indian Cucurbitaceae. *Bull. Bot. Soc. Bengal* 15(1 & 2): 4-16.

Chakravarty, H.L. 1966. Monograph on the Cucurbitaceae of Iraq. *Iraq Minist. Agric. Techn. Bull.* 3: 1-145.

Chakravarty, H.L. 1968. Cucurbitaceae of Ghana. *Bull. Inst. Fondam. Afrique Noire, sér. A, Sci. Nat.* 18(2): 400-408.

Chakravarty, H.L., and C. Jeffrey. 1980. Cucurbitaceae. In *Fl. Iraq*, eds. C.C. Townsend and E. Guest, 4(1): 191-208. Bagdad: Ministry of Agriculture & Agrarian Reform.

Chatterjee, M., and T.A. More. 1991. Interspecific hybridization in *Cucumis* spp. *Cucurbit Genetics Cooperative* 14: 69-70.

Chelliah, S., and C.N. Sambandam. 1972. Inheritance of resistance to the fruit fly, *Dacus cucurbitae* C., in the interspecific cross between *Cucumis callosus* (Rottl.) Cogn. and *Cucumis melo* L. *Auara* 4/5: 169-171.

Chiovenda, E. 1929. *Fl. somala*, vol. 1. Rome: Pubblicazione a cura del ministero delle colonie.

Chopra, R.H. 1955. Some observations on endosperm development in the Cucurbitaceae. *Phytomorphology* 5: 219-230.

Clarke, C.B. 1879. Cucurbitaceae. In *Fl. Brit. India*, ed. J.D. Hooker, 2: 604-635. London: Reeve & Co.

Cogniaux, A. 1881. Cucurbitaceae. In *Monogr. phan.*, eds. A.L.P.P. de Candolle and A.C.P. de Candolle, 3: 325-951, 953-954. Paris: G. Masson.

Cogniaux, A. and H. Harms. 1916. Cucurbitaceae—Fevilleae et Melothrieae. In *Pflanzenr.*, ed. A. Engler, heft 66. Leipzig: Wilhelm Engelmann.

Cogniaux, A. and H. Harms. 1924. Cucurbitaceae—Cucurbiteae—Cucumerinae. In *Pflanzenr.*, ed. A. Engler, heft 88. Leipzig: Wilhelm Engelmann.

Cufodontis, G. 1965. Enumeratio plantarum Aethiopae spermatophyta, 2. *Bull. Jard. Bot. État., suppl.*, 35(2): 1011-1058.

Dallwitz, M.J. 1974. A flexible computer program for generating identification keys. *Syst. Zool.* 23: 50-57.

Dallwitz, M.J. 1991. Draft definition of the DELTA format. *DELTA Newslett.* 7: 1-6.

Dallwitz, M.J., and T.A. Paine. 1986. *User's guide to the DELTA system - a general system for processing taxonomic descriptions*, ed. 3. CSIRO Austral. Div. Entomol. Rep. 13: 1-106.

Dane, F. 1976. *Evolutionary studies in the genus* Cucumis. Unpubl. Dissertation, Colorado State University, Fort Collins.

Dane, F. 1983. Cucurbits. In *Pl. Genet. Breed. pt. B*, pp. 369-390. Amsterdam: Elsevier Sci. Publ.

Dane, F., D.W. Denna, and T. Tsuchiya. 1980. Evolutionary studies of wild species in the genus *Cucumis*. *Z. Pflanzenzücht.* 85: 89-109.

Dane, F., and T. Tsuchiya. 1976. Chromosome studies in the genus *Cucumis*. *Euphytica* 25: 367-374.

Dane, F., and T. Tsuchiya. 1979. Meiotic chromosome and pollen studies of polyploid *Cucumis* species. *Euphytica* 28: 563-567.

Davis, P.H., and V.H. Heywood. 1973. *Principles of angiosperm taxonomy*. Huntington: Robert E. Krieber Publ. Co.

Davy, J.B. 1926. *Man. pl. Transvaal*. London: Longmans, Green & Co.

Deakin, J.R., G.W. Bohn, and T.W. Whitaker. 1971. Interspecific hybridization in *Cucumis*. *Econ. Bot.* 25(2): 195-211.

Delile, A.R. 1843. *Index seminum horti regii botanici Monspeliensis, anni 1843.* Montpellier: J. Martel.

Delile, A.R. 1847. *Index seminum horti regii botanici Monspeliensis, anni 1847.* Montpellier: J. Martel.

Dodoens, R. 1583. *Stirp. hist. pempt.* Antwerp: officina Christophori Plantini.

Doebley, J.F., and H.H. Iltis. 1980. Taxonomy of *Zea* (Graminae). I. A subgeneric classification with key to taxa. *Amer. J. Bot.* 67(6): 982-993.

Enslin, P.R. and S. Rehm. 1957. The distribution and biogenesis of the cucurbitacins in relation to the taxonomy of the Cucurbitaceae. *Proc. Linn. Soc. London* 168: 230-238.

Esquinas-Alcazar, J.T. 1977. *Alloenzyme variation and relationships in the genus* Cucumis. Unpubl. Dissertation, University of California, Davis.

Fassuliotis, G. 1977. Self-fertilization of *Cucumis metuliferus* Naud. and its cross-compatilibity with *C. melo* L. *J. Amer. Hort. Soc.* 102(3): 336-339.

Feinbrun-Dothan, N. 1978. Cucurbitaceae. In *Fl. palaestina* 3(text): 271-276; 3(plates): 471-472. Jerusalem: Israel Academy of Sciences and Humanities.

Fernandes, R., and A. Fernandes. 1962. Cucurbitaceae de Angola. *Mem. Junta Invest. Ultram., sér. 2* 34: 29-150.

Fernandes, R., and A. Fernandes. 1970. 88—Cucurbitaceae. In *Consp. fl. angolensis*, eds. A.W. Exell, A. Fernandes, and E.J. Mendes, 4: 232-289. Lisbon: Junta de Investigações do Ultramar e Instituto de Investigações Científica de Angola.

Fernandes, R.B. 1976. Cucurbitaceae de Cabo Verde. *Garcia de Orta*, sér. bot. 3(1): 1-14.

Fernandes, R.B. 1985. Éclaircissement sur l'identification d'une espèce de *Cucumis* des Iles de Cap Vert. *Bol. Soc. Brot.*, sér. 2, 58: 219-234.

Filov, A.I. 1960. The problem of melon systematics. *Vestn. Skh. Nauki* 1: 126-132. [in Russian].

Fursa, T.B., and I. Filov. 1982. XXI Cucurbitaceae (*Citrullus, Cucurbita*). In *Flora of Cultivated Plants*, eds. O.N. Korovina and T.B. Fursa, vol 21. Moscow: Kolos.

Gabaev, S.G. 1929. Cucumbers of Asia. *Trudy Prikl. Bot.* 23(3): 443-473. [in Russian, and translated into English for USDA by G. Saad in 1986]

Gabaev, S.G. 1932. *Cucumber*, 212 pp. Leningrad: Institut rastenievodstva NKZ SSSR. [in Russian, and translated into English for USDA by G. Saad in 1987]

Gabaev, S.G. 1933a. Systematische Untersuchungen an Gurkenarten und Varietäten. *Angew. Bot.* 15: 290-307. [in German].

Gabaev, S.G. 1933b. Cucumbers. In *La Turquie Agricole*, ed. P. Zhokovsky, 552-559, 880-882. Moscow: State Agriculture Section "Selkhozghiz." [in Russian with extensive summary in French].

Gibbs Russell, G.E. 1985. PRECIS: The National Herbarium's computerised information system. *S. African J. Sci.* 81(2): 62-65.

Gibbs Russell, G.E., W.G. Welman, E. Retief, K.L. Immelman, G. Germishuizen, B.J. Pienaar, M. van Wyk, A. Nicolas, C. de Wet, J.C. Mogford, and J. Mulvenna. 1987. List of species of southern African plants, ed. 2, recent literature and synonyms, part 2, dicotyledons. *Mem. Bot. Surv. South Africa* 56: 1-270.

Grebenscikov, I. 1953. Die Entwicklung del Melonsystematik (Ein Beitrag zur Systematik der Kulturpflanzen). *Kulturpflanzen* 1: 121-138. [in German].

Greuter, W., et al. 1988. *International code of botanical nomenclature.* Königstein: Koetlz Scientific Books.

Gulyayev, V.A. 1963. Comparative embryology of the Cucurbitaceae and its significance for the systematics of the family. *Bot. Zhurn.* (*Moscow & Leningrad*) 48: 80-84.

96 Literature Cited

Gunn, M. and L.E. Codd. 1981. *Botanical exploration of southern Africa.* Cape Town: A.A. Balkema.

Halliday, P. and C. Jeffrey. 1978. *Cucumis.* In *Fl. Zambesiaca,* ed. E. Launert, 4: 462-475.

Harrison, S. 1978. Recent work on the herbarium of John Forbes Royle (1799-1858) at Merseyside County Museum (LIV). *Taxon* 27(1): 21-33.

Harz, C.K.O. 1885. Landw. Samenk. Berlin: Paul Parey.

Hassib, M. 1938. *Cucurbit. Egypt, Fouad Univ.,* publ. no. 3. Cairo: F.E. Noury & fils.

Heiser, C.B., E.E. Schilling, and B. Dutt. The american species of *Luffa* (Cucurbitaceae). *Syst. Bot.* 13(1): 138-145. 1988.

Hiern, W.P. 1896-1898. *Cat. afr. pl.,* vol. 1. London: British Museum.

Holmgren, P.K., N.H. Holmgren, and L.C. Barnett. 1990. *Index herbariorum. Part I: The Herbaria of the World.* Bronx: New York Botanical Garden.

Hooker, J.D. 1871. Cucurbitaceae. In *Flora of Tropical Africa,* ed. D. Oliver, 2: 521-569.

Howard, R.A. 1973. The Enumeratio and Selectarum of Nicolaus von Jacquin. *J. Arnold Arbor.* 54(4): 435-470.

Iltis, H.H., and J.F. Doebley. 1980. Taxonomy of *Zea* (Graminae). I. Subspecific categories in the *Zea mays* complex and a generic synopsis. *Amer. J. Bot.* 67(6): 994-1004.

Inamdar, J.A., and M. Gangadhara. 1975. Structure, ontogeny, classification and organographic distribution of trichomes in some Cucurbitaceae. *Feddes Repert.* 86: 307-320.

Inamdar, J.A., M. Gangadhara, and K.N. Shenoy. 1990. Structure, ontogeny, distribution, and taxonomic significance of trichomes and stomata in the Cucurbitaceae. In *Biology and utilization of the Cucurbitaceae,* eds. D.M. Bates, R.W. Robinson, and C. Jeffrey, pp. 209-224. Ithaca: Cornell University Press.

Jackson, B.D. 1928. *Gloss. bot. terms* ed. 4. London: Gerald Duckworth & Co.

Jirásek, V. 1961. Evolution of the proposals of taxonomic categories for the classification of cultivated plants. *Taxon* 10(2): 34-45.

Jeffrey, C. Mar 1962. Notes on Cucurbitaceae, including a proposed new classification of the family. *Kew Bull.* 15(3): 337-371.

Jeffrey, C. 1964. A note on pollen morphology in Cucurbitaceae. *Kew Bull.* 17(3): 473-477.

Jeffrey, C. 1965. Further notes on Cucurbitaceae. *Kew Bull.* 19(2): 215-223.

Jeffrey, C. 1967a. Cucurbitaceae. In *Fl. Trop. East Africa,* eds. E. Milne-Redhead and R.M. Polhill. Crown Agents for Oversea Governments and Adminstrations.

Jeffrey, C. 1967b. On the classification of the Cucurbitaceae. *Kew Bull.* 20(3): 417-426.

Jeffrey, C. 1975. Further notes on Cucurbitaceae: III. Some southern African taxa. *Kew Bull.* 30(3): 475-493.

Jeffrey, C. 1978. *Cucumella.* In *Fl. Zambesiaca,* ed. E. Launert, 4: 472-475.

Jeffrey, C. 1980a. A review of the Cucurbitaceae. *J. Linn. Soc., Bot.* 81: 233-247.

Jeffrey, C. 1980b. Further notes on Cucurbitaceae: V. The Cucurbitaceae of the Indian subcontinent. *Kew Bull.* 34(4): 789-809.

Jeffrey, C. 1980c. *A report on studies made during, and in connexion with, a visit to China, Hong Kong and Japan in 1980.* Kew: Royal Botanic Gardens.

Jeffrey, C. 1985. Further notes on Cucurbitaceae: VII Preliminary to the Flora of Ethiopia. *Kew Bull.* 40(1): 209-211.

Jeffrey, C. 1990. Systematics of the Cucurbitaceae: An overview. In *Biology and utilization of the Cucurbitaceae,* eds. D.M. Bates, R.W. Robinson, and C. Jeffrey, pp. 3-9, 449-463. Ithaca: Cornell University Press.

Jones, W.H.S., transl. 1951. *Pliny, natural history,* vol. 6, books 20-23. Cambridge: Harvard Univ. Press.

Keraudren, M. 1966. 185e Famille.—Cucurbitacées. In *Fl. Madagascar Comores (Pl. vasc.)*, ed. H. Humbert, 173 pp. Paris: Muséum National d'Histoire Naturelle.

Keraudren, M. 1967. Cucurbitacées. In *Fl. Cameroun*, ed. A. Aubreville, 6: 1-192. Paris: Muséum National d'Histoire Naturelle.

Keraudren, M. 1968. Recherches sur les Cucurbitacées de Madagascar. *Mém. Mus. Natl. Hist. Nat., sér. B, Bot.* 16(2): 127-330, *pl. 1-13*.

Keraudren-Aymonin, M. 1975. Cucurbitaceae. In *Fl. Afr. Centr. (Zaïre-Rwanda-Burundi)*, ed. P. Bamps. Bruxelles: Jardin botanique national de Belgique.

Kho, Y.O., A.P.M. den Nijs, and J. Franken. 1980. Interspecific hybridization in *Cucumis* L. II. The crossability of species, an investigation of in vivo pollen tube growth and seed set. *Euphytica* 29: 661-672.

Kirkbride, J.H., Jr. 1992a. Database of native distribution for *Cucumis*. *Cucurbit Genetics Cooperative* 15: 45-47, 5¼ inch 1.2 megabyte diskette.

Kirkbride, J.H., Jr. 1992b. Interactive microcomputer database for identification of *Cucumis*. *Cucurbit Genetics Cooperative* 15: 48-50, 5¼ inch 1.2 megabyte diskette.

Knerr, L.D., and J.E. Staub. 1991. A multivariate re-evaluation of biochemical genetic diversity in *Cucumis sativus* L. *Cucurbit Genetics Cooperative* 14: 25-28.

Kozuchov, Z.A. 1930. Karyological investigations of the genus *Cucumis*. *Trudy Prikl. Bot.* 23: 357-365.

Lawrence, G.H.M. 1955. *An introduction to plant taxonomy*. New York: The Macmillan Co.

Leeuwen, L. van, and A.P.M. den Nijs. 1980. Problems with the identification of *Cucumis* L. taxa. *Cucurbit Genetics Cooperative* 3: 55-59.

Linnaeus, C. 1738. *Hortus cliffortianus*. Amsterdam: Amstelaedami.

Linnaeus, C. 1753. *Sp. pl. ed. 1*. Stockholm: Impensis Laurentii Salvii.

Linnaeus, C. 1754. *Gen. pl. ed. 5*. Stockholm: Impensis Laurentii Salvii.

Lu, A.-M. and Z.-Y. Zhang. 1986. Cucurbitaceae. In *Flora Republicae Popularis Sinicae*, eds. A.-M. Lu and S.-K. Chen, vol. 73(1). Beijing: Science Press.

Mallick, M.F.R., and M. Masui. 1986. Origin, distribution and taxonomy of melons. *Scientia Hort.* 28(3): 251-261.

Manniche, L. 1989. *An ancient Egyptian herbal*. Austin: Univ. Texas Press.

Marticorena, C. 1963. Material para una monografia de la morfología del polen de Cucurbitaceae. *Grana Palynol.* 4: 78-91.

Merrill, E.D. 1917. *An interpretation of Rumphius's Herbarium amboinense*, 595 pp. Manila: Bureau of Printing.

Meeuse, A.D.J. 1955. The aardvark cucumber. *Farming South Africa* 30(351): 301-304.

Meeuse, A.D.J. 1958. A possible case of interdependence between a mammal and a higher plant. *Arch. Néerl. Zool.* 13, suppl.: 314-318.

Meeuse, A.D.J. May 1962. The Cucurbitaceae of southern Africa. *Bothalia* 8(1): 1-111.

Morris, J.W., and H.F. Glen. 1978. PRECIS, the National Herbarium of South Africa (PRE) computerized information system. *Taxon* 27(5/6): 449-462.

Munger, H.M., and R.W. Robinson. 1991. Nomenclature of *Cucumis melo* L. *Cucurbit Genetics Cooperative* 14: 43-44.

Naudin, C.V. 1859. Essais d'une monographie des espèces et des variétés du genre *Cucumis*. *Ann. Sci. Nat. Bot., sér.* 4, 11: 5-87.

Naudin, C.V. 1862. Cucurbitacées cultivées au Muséum d'Histoire Naturelle en 1862. *Ann. Sci. Nat. Bot., sér.* 4, 18: 159-208.

Naudin, C.V. 1866. Cucurbitacées nouvelles cultivées au Muséum d'Histoire Naturelle en 1863, 1864 et 1865. *Ann. Sci. Nat. Bot., sér.* 5, 5: 5-43.

Naithani, S.P. 1947. Somatic chromosome numbers in some cultivated cucurbits. *Curr. Sci.* 16(6): 188-189.

Nazimuddin, S., and S.S.H. Naqvi. 1984. No. 154 Cucurbitaceae. In *Fl. Pakistan*, eds. E. Nasir and S.I. Ali, 154: 1-56. Karachi: Univ. of Karachi.

Nei, M. 1972. Genetic distance between populations. *Amer. Naturalist* 106: 283-292.

Neier, R., and C. Marr. 1988. Vegetable preferences of home gardeners. *Hortscience* 23(5): 916.

Nicolson, D.H., C.R. Suresh, and K.S. Manilal. 1988. An interpretation of Van Rheede's Hortus Malabaricus. *Regnum Veg.* 119: 1-378.

Nijs, A.P.M. den, and J.B.M. Custers. 1990. Introducing resistances into cucumbers by interspecifc hybridization. In *Biology and utilization of the Cucurbitaceae*, eds. D.M. Bates, R.W. Robinson, and C. Jeffrey, pp. 382-396. Ithaca: Cornell University Press.

Nijs, A.P.M. den, and D.L. Visser. 1985. Relationships between African species of the genus *Cucumis* L. estimated by the production, vigour and fertility of F_1 hybrids. *Euphytica* 34: 279-290.

Pangalo, K.J. 1925. Turkestan melons. *Trudy Prikl. Bot. Selekc.* 14(1): 37-70. [in Russian, and translated into English for USDA by G. Saad in 1987].

Pangalo, K.J. 1928a. *Melons*, 95 pp. Leningrad: Izdanie Vsesoyuznogo Institute Prikladnoi Botaniki i Novykh Kultur. [in Russian, and translated into English for USDA by G. Saad in 1986].

Pangalo, K.J. 1928b. Studien über die Systematik und Geographie der Melonen. *Deutsche Landw. Rundschau* 2: 61-67. [in German].

Pangalo, K.J. 1929a. Varieties of Russian musk melon, their composition and origin. *Trudy Prikl. Bot.* 23(3): 21-40. [in Russian, and translated into English for USDA by G. Saad in 1986]

Pangalo, K.J. 1929b. Cantaloupes, their history and new prospects of improvement. *Trudy Prikl. Bot.* 23(3): 209-228. [in Russian, and translated into English for USDA by G. Saad in 1986]

Pangalo, K.J. 1929c. Wild melons. *Trudy Prikl. Bot.* 23(3): 231-252. [in Russian, and translated into English for USDA by G. Saad in 1986]

Pangalo, K.J. 1929d. Critical review of the main literature on the taxonomy, geography and origin of cultivated and partially wild melons. *Trudy Prikl. Bot.* 23(3): 397-442. [in Russian, and translated into English for USDA by G. Saad in 1986]

Pangalo, K.J. 1929e. Chinese melons. *Trudy Prikl. Bot.* 23(3): 545-560. [in Russian, and translated into English for USDA by G. Saad in 1986]

Pangalo, K.J. 1933a. *Melon cultivation in the USSR*, 224 pp. Moscow-Leningrad: OGIZ. [in Russian, and translated into English for USDA by G. Saad in 1987]

Pangalo, K.J. 1933b. Melons. In *La Turquie Agricole*, ed. P. Zhokovsky, 520-543, 874-877. Moscow: State Agriculture Section "Selkhozghiz." [in Russian with extensive summary in French].

Pangalo, K.J. 1934. *Key to the varieties of melo crops*, 56 pp. Leningrad: Vsesoiuznyi Institut. [in Russian, and translated into English for USDA by G. Saad in 1986]

Pangalo, K.J. 1937. Breeding of melon crops. In *Theor. bases pl. breeding*, ed. N.I. Vavilov, 3: 135-194. [in Russian, and translated into English for USDA by G. Saad in 1987].

Pangalo, K.J. 1950. Melons as the independent genus *Melo* Adans. *Bot. Zhurn.* (*Moscow & Leningrad*) 35(6): 571-580. [in Russian].

Payne, W.W. 1978. A glossary of plant hair terminology. *Brittonia* 30(2): 239-255.

Partridge, T.R., M.J. Dallwitz, and L. Watson. 1991. A Primer for the DELTA System on MS-DOS and VMS, ed. 2.2. *CSIRO Austral. Div. Entomol. Rep.* 38: 1-20.

Pas, H.N. ten, J.W.P. Schoenaker, E.H. Oost, and C.E. Jarvis. 1985. Re-lectotypification of *Cucumis sativus* L. *Taxon* 34(2): 288-293.

Perl-Treves, R. 1985. *The plastome of* Cucumis: *Phylogenetic relationships expressed by the chloroplast DNA restriction patterns.* Unpubl. Thesis, Weizmann Institute of Science, Rehovot.

Perl-Treves, R. and E. Galun. 1985. The *Cucumis* plastome: physical map, intrageneric variation and phylogenetic relationships. *Theor. Appl. Genet.* 71: 417-429.

Perl-Treves, R., D. Zamir, N. Navot, and E. Galun. 1985. Phylogeny of *Cucumis* based on isozyme variablity and its comparison with plastome phylogeny. *Theor. Appl. Genet.* 71: 430-436.

Perrino, P., K. Hammer, and P. Hanelt. 1986. Carosello and the taxonomy of *Cucumis melo* L., especially its vegetable races. *Acta Hort.* 182: 95-100.

Polhill, D. 1988. Index of collecting localities. In *Flora of Tropical East Africa*, eds. E. Milne-Redhead and R.M. Polhill. Crown Agents for Oversea Governments and Adminstrations.

Public Buildings Office. *World wide geographic location codes.* Washington, D.C.: U.S. General Services Administration.

Puchalski, J.T., and R.W. Robinson. 1990. Electrophoretic analysis of isozymes in *Cucurbita* and *Cucumis* and its application for phylogenetic studies. In *Biology and utilization of the Cucurbitaceae*, eds. D.M. Bates, R.W. Robinson, and C. Jeffrey, pp. 60-76. Ithaca: Cornell University Press.

Puchalski, J.T., R.W. Robinson, and J.W. Shail. 1978. Comparative electrophoresis of isozymes of *Cucumis* species. *Cucurbit Genetics Cooperative* 1: 39.

Qi, C.Zh., Zh.Zh. Yuan, and Y.X. Li. 1983. A new type of cucumber--*Cucumis sativus* L. var. *xishuangbannanesis* [in Chinese]. *Acta Hort. Sin.* 10(4): 259-263.

Raamsdonk, L.W.D., A.P.M. den Nijs, and M.C. Jongerius. 1989. Meiotic analyses of *Cucumis* hybrids and an evolutionary evaluation of the genus *Cucumis* (Cucurbitaceae. *Pl. Syst. Evol.* 163: 133-146.

Raamsdonk, L.W.D., and D.L. Visser. 1992. Autotetraploidy in *Cucumis zeyheri* and a derived new species *C. diniae* (Cucurbitaceae). *Nord. J. Bot.* 12(3): 327-334.

Radford, A.E., et al. 1974. Vasc. pl. syst. New York: Harper & Row.

Rackham, H., transl. 1950. *Pliny, natural history*, vol. 5, books 17-19. Cambridge: Harvard Univ. Press.

Ramachandran, C. 1984. Nuclear DNA variation in *Cucumis* species. *Cucurbit Genetics Cooperative* 7: 97-98.

Ramachandran, C., and R.K.J. Narayan. 1985. Evolution of chromosomal DNA in *Cucumis*. *Theor. Appl. Gen.* 69(5/6): 497-502.

Ramachandran, C., and V.S. Seshadri. 1986. Cytological analysis of the genome of cucumber (*Cucucmis sativus* L.) and muskmelon (*Cucumis melo* L.). *Z. Pflanzenzücht.* 96: 25-38.

Reekmans, M. 1983. Cucurbitaceae. In *Fl. Rwanda*, ed. G. Tropin, 2: 453-480.

Rehm, S. 1960. Die Bitterstoffe der Cucurbitaceen. *Ergebn. Biol.* 22: 108-136.

Rehm, S., P.R. Enslin, A.D.J. Meeuse, and J.H. Wessels. 1960. Bitter principles of the Cucurbitaceae. VII.--The distribution of bitter principles in this plant family. *J. Sci. Food Agric.* 8: 679-686.

Richard, A. 1847. *Tent. fl. abyss.*, vol. 1. Paris: Arthus Bertrand.

Roessler, H. 1968. 94. Cucurbitaceae. In *Prodr. Fl. Südwestafrika*, ed. H. Merxmüller. Lehre: Cramer.

Ruitter, A.-C. de. 1974. Crisement interspécifique entre concombre (*Cucumis sativus*) et melon (*Cucumis melo*). *Agron. Trop.* (*Nogent-sur-Marne*) 29: 845.

Schrader, H.A. 1838. Reliquiae Schraderianae. *Linnaea* 12: 353-476.

Schuman, D.A., J.E. Staub, and B.E. Struckmeyer. 1985. Morphological and anatomical comparisons between two *Cucumis sativus*, botanical varieties: *hardwickii* and *sativus*. *Cucurbit Genetics Cooperative* 8: 15-18.

Shimotsuma, M. 1965. Chromosome studies of some *Cucumis* species. *Rep. Kihara Inst. Biol. Res.* 17: 11-16.

Singh, A.K. 1990. Cytogenetics and evolution in the Cucurbitaceae. In *Biology and utilization of the Cucurbitaceae*, eds. D.M. Bates, R.W. Robinson, and C. Jeffrey, pp. 10-28. Ithaca: Cornell University Press.

Singh, A.K., and R.P. Roy. 1974. Karyological studies in *Cucumis* (L.). *Caryologia* 27(2): 153-160.

Singh, A.K., and R.P. Roy. 1975. Caryological studies in *Cucumis hardwickii* (Royal). *J. Indian Bot.* 54: 94-97.

Singh, A.K., and K.S. Yadava. 1984. An analysis of interspecific hybrids and phylogenetic implications in *Cucumis* (Cucurbitaceae). *Pl. Syst. Evol.* 147: 237-252.

Singh, D., and A.S.R. Dathan. 1974. Structure and development of seed coat, *Cucumis* L. *New Bot.* 1(1+2): 8-22.

Sonder, O.W. 1862. Cucurbitaceae. In *Fl. cap.*, eds. W.H. Harvey and O.W. Sonder, 2: 482-498. London: Reeve & Co.

Spix, J.B. von, and C.F.P. von Martius. 1823-1831. *Reise Bras.*, 3 vols. München: M. Lindauer/I.J. Lentner/Friedr. Fleischer.

Stace, C.A. 1980. *Plant taxonomy and biosystematics*. Baltimore: Univ. Park Press.

Stafleu, F.A., et al. 1978. *International Code of Botanical Nomenclature*. Utrecht: Bohn, Scheltema & Holkema.

Stafleu, F.A. and R.S. Cowan 1981. *Taxonomic Literature*, vol 3, Lh-O. Utrecht: Bohn, Scheltema & Holkema.

Staub, J.E., L. Fredrick, and T. Marty. 1987. Electrophoretic variation in cross-compatible wild diploid species of *Cucumis*. *Canad. J. Bot.* 65(4): 792-798.

Staub, J.E., L.D. Knerr, D.J. Holder, and B. May. 1991. Phylogenetic relationships among several African *Cucumis* species. *Canad. J. Bot.*, in press.

Staub, J.E., and R.S. Kupper. 1986. Discrimination among seven *Cucumis hardwickii* (R) Alef. accessions based on principal component analysis. *Cucurbit Genetics Cooperative* 9: 27-29.

Stebbins, G.L. 1971. *Chromosomal evolution in higher plants*. London: Edward Arnold Ltd.

Stearn, W.T. 1966. *Bot. Latin*. New York: Hafner Publ. Co.

Swart, J.J. 1979. *Melo*. In *Index nominum genericorum (plantarum)*, eds. E.R. Farr, J.A. Leussink, and F.A. Stafleu, vol. 2. Utrecht: Bohn, Scheltema & Holkema.

Systematic Association Committee for Descriptive Terminology. 1962. Terminology of simple symmetrical plane shapes (chart 1). *Taxon* 11(5): 145-156, *chart 1*.

Telford, I.R. 1982. Cucurbitaceae. In *Flora of Australia*, ed. B.G. Briggs, 8: 158-198. Canberra: Australian Government Publishing Service.

Thomas, C.E., ed. 1989. *Proceedings of Cucurbitaceae 89: Evaluation and enhancement of Cucurbit germplasm, November 29-December 2, 1989, Charleston, SC*. Charleston: U.S. Vegetable Laboratory.

Thulin, M. 1991. New species of *Cucumis* (Cucurbitaceae) from northeast tropical Adrica. *Nord. J. Bot.* 11(5): 535-542. 1991

Trivedi, R.N., and R.P. Roy. 1970. Cytological studies in *Cucumis* and *Citrullus*. *Cytologia* 35: 561-569.

Varekamp, H.Q., D.L. Visser, and A.P.M. den Nijs. 1982. Rectification of the names of certain accessions of the IVT-*Cucumis* collection. *Cucurbit Genetics Cooperative* 5: 59-60.

Verheyen, R. 1951. *Contribution à l'étude éthologique des mammifères du Parc National de l'Upemba.* Bruxelles: Institut des Parcs Nationaux du Congo Belge.

Visser, D.L., and A.P.M. den Nijs. Variation for interspecific crossability of *Cucumis anguria* L. and *C. zeyher* Sond. *Cucurbit Genetics Cooperative* 6: 100-101.

Voss, E.G., et al. 1983. *International Code of Botanical Nomenclature.* Utrecht: Bohn, Scheltema & Holkema.

Watt, J.M., and M.G. Breyer-Brandwijk. 1962. *Med. poison. pl. S. E. Afr. ed. 2.* Edinburgh: E. & S. Livingston.

Whitaker, T.W. 1933. Cytological and phylogenetical studies in the Cucurbitaceae. *Bot. Gaz.* (*Crawfordsville*) 94: 780-790.

Whitaker, T.W., and G.N. Davis. 1962. *Cucurbits, botany, cultivation, and utilization.* New York: Interscience Publishers.

White, F. 1983. *The vegetation of Africa, a descriptive memoir to accompany the Unesco/AETFAT/UNSO vegetation map of Africa.* Paris: Unesco.

Woodward, M., ed. 1964. *Leaves from Gerard's herball.* London: Spring Books.

Yadava, K.S., A.K. Singh, and H.C. Arya. 1984. Cytogenetic investigation in *Cucumis* L. I. Meiotic analysis in twenty four *Cucumis* species. *Cytologia* 49: 1-9.

APPENDIXES

Appendix 1. Synonyms of *Cucumis melo* Linnaeus

Bryonia callosa Wallich, Numer. List no. 6710. 1832. Nom. nud.
Bryonia collosa Rottler, Ges. Naturf. Freunde Berlin Neue Schriften 4: 210. 1803.—TYPE: INDIA. Madras, *J.P. Rottler s.n.* (holotype: K!).
Cucumis acidus N.J. Jacquin, Observ. Bot. 4: 14. 1771.
Cucumis agrestis (Naudin) Grebenscikov, Kulturpflanze 1: 134. 1953. Invalid, lacks full and direct reference to place of valid publication of its basionym.
Cucumis agrestis (Naudin) Grebenscikov subsp. *figari* (Pangalo) Grebenscikov, Kulturpflanze 1: 134. 1953.—BASIONYM: *Melo figari* Pangalo. Invalid, its basionym is invalid.
Cucumis alba Nakai, Rigakukai 13(9): 14. 1916.
Cucumis alba Nakai var. *flavescens* Nakai, Rigakukai 13(9): 15. 1916.
Cucumis ambigua Fenzl ex Hooker filius in Oliver, Fl. trop. Afr. 2: 543. 1871.—TYPE: SUDAN. Sennar: Grujeschal, 24 Mar 1840, *K.G.T. Kotschy 352* (BM! BR! E! L! LE! M!). Invalid, published as synonym of *C. dipsaceus.*
Cucumis arenarius Schumacher & Thonning, Beskr. Guin. pl. 200. 1827.
Cucumis aromaticus Royle, Illustr. Bot. Himal. 2: 220. 1835.—TYPES: INDIA. Saharunpore, *Royle 80/15* (syntype: LIV!); *Royle S306* (syntype: LIV!). Invalid, not accepted by its author.
Cucumis bardana Fenzl ex Naud., Ann. Sc. Nat. Sér. 4, 11: 25. 1859. Invalid, published as synonym of *C. dipsaceus.*
Cucumis bisexualis A.-M. Lu & G. C. Wang ex Lu & Z.-Y. Zhang, Bull. Bot. Res. North-East Forest. Inst. 4(2): 126. 1984.—China, *G.C. Wang sn.*
Cucumis bucharicus Hortulanorum ex Steudel, Nomencl. 451. 1840. Invalid, published as synonym of *C. melo.*
Cucumis campechianus Kunth in H.B.K., Nov. gen. sp. 1: 122 quarto, 97 folio. 1816.—TYPE: MEXICO. Campeche, *F.W.H.A. von Humbodlt & A.J.A. Bonpland s.n.* (holotype: P-Bonpl.!).
Cucumis cantalou Hortulanorum ex Steudel, Nomencl. 451. 1840. Invalid, published as synonym of *C. melo.*
Cucumis cantalupensis Haberle ex M. Roemer, Fam. nat. syn. monogr. 2: 69. 1846. Illegitimate, *Cucumis cantalupo* Haberle ex H.G.L. Reichenbach given in synonymy.
Cucumis cantalupo Haberle ex H.G.L. Reichenbach, Fl. Germ. Excurs. 295. 1831.
Cucumis chate Hasselquist, Iter. Palest. 491. 1757.—TYPE: SWEDEN. Plant cultivated at Uppsala, Sweden (neotype: LINN, sheet number 1152.11! [Jeffrey, 1980]). Jeffrey (1980) cited the Species plantarum (Linnaeus, 1753) number--1152/5--not the LINN number--1152.11.
Cucumis chinensis Pangalo, Bull. Applied Bot., Lenin., 1929-30, 23(3): 547. 1930. Illegitimate, *Cucumis conomon* Thunberg given in synonymy.
Cucumis chito Morr., Ann. Soc. Gand. 5: 341. 1849.—TYPE (here designated): *T. 278, Ann. Soc. Gand.*, vol. 5 (lectotype).
Cucumis cicatrisatus Stocks, Hook. Kew Journ. 4: 148. 1852.—PAKISTAN. Scinde, *J.E. Stocks s.n.* (holotype: K).
Cucumis cognatus Fenzl ex Hooker filius in Oliver, Fl. trop. Afr. 2: 543. 1871. Invalid, not accepted by its author.
Cucumis collosus (Rottler) Cogniaux in Cogniaux & Harms, Pflanzenr. 88: 129. 1924.—BASIONYM: *Bryonia collosa* Rottler.
Cucumis conomon Thunberg, Nova Acta Soc. Sc. Upsal. 3: 208. 1780.
Cucumis conomon Thunberg var. *koreana* Nakai, Rigakukai 13(9): 14. 1916.
Cucumis conomon Thunberg var. *rugosa* Nakai in T. Mori, Enum. pl. Corea 335. 1922. Nom. nud.
Cucumis cubensis Schrader, Linnaea 12: 419. 1838.
Cucumis cubensis Schrader var. *costatus* M. Roemer, Fam. nat. syn. monogr. 2(1): 76. 1846.
Cucumis deliciosus R.A. Salisbury, Prodr. stirp. Chap. Allerton 157. 1796. Illegitimate, *C. melo* Linnaues given in synonymy.

Cucumis dudaim Linnaeus, Sp. pl. *ed. 1*, 1011. 1753.—TYPE: Dillenius, Hort. eltham. 1: *t. 177, f. 218*. 1732 (lectotype [Jeffrey, 1980]).

Cucumis dudaim Linnaeus *aegyptiaca* Sickenberger, Mem. Ins. Egypt. 4(2): 232. 1901. Published without indication of rank, so inoperative for priority except for homonymy.

Cucumis dudaim Linnaeus *bitariana* Sickenberger, Mem. Ins. Egypt. 4(2): 231. 1901. Published without indication of rank, so inoperative for priority except for homonymy.

Cucumis dudaim Linnaeus *elongata* Sickenberger, Mem. Ins. Egypt. 4(2): 233. 1901. Published without indication of rank, so inoperative for priority except for homonymy.

Cucumis dudaim Linnaeus *striata* Sickenberger, Mem. Ins. Egypt. 4(2): 232. 1901. Published without indication of rank, so inoperative for priority except for homonymy.

Cucumis eriocarpus Boissier & Noé in Boissier, Diagn. pl. orient., sér. 2, 2: 59. 1856.—TYPE: IRAQI. In cultis propè Bagdad, *Noé s.n.* (holotype: G).

Cucumis erivanicus Hortulanorum ex Steudel, Nom. *ed.* 2, 1: 451. 1840. Invalid, published as synonym of *C. melo* Linnaeus.

Cucumis eumelo Pangalo in Vavilov, Theoret. bases pl. breed. 3: 147. 1937. Orthographic variant of *Cucumis melo* Linnaeus.

Cucumis eumelo subsp. *adana* Pangalo in Vavilov, Theoret. bases pl. breed. 3: 148. 1937. Invalid, published as synonym of *Cucumis melo* Linnaeus.

Cucumis eumelo subsp. *ameri* Panaglo in Vavilov, Theoret. bases pl. breed. 3: 149. 1937. Invalid, published as synonym of *Cucumis melo* Linnaeus.

Cucumis eumelo subsp. *cantalupa* Pangalo in Vavilov, Theoret. bases pl. breed. 3: 148. 1937. Invalid, published as synonym of *Cucumis melo* Linnaeus.

Cucumis eumelo subsp. *cassaba* Panaglo in Vavilov, Theoret. bases pl. breed. 3: 148. 1937. Invalid, published as synonym of *Cucumis melo* Linnaeus.

Cucumis eumelo subsp. *chandaljac* Pangalo in Vavilov, Theoret. bases pl. breed. 3: 149. 1937. Invalid, published as synonym of *Cucumis melo* Linnaeus.

Cucumis eumelo subsp. *zard* Pangalo in Vavilov, Theoret. bases pl. breed. 3: 149. 1937. Invalid, published as synonym of *Cucumis melo* Linnaeus.

Cucumis flexuosus Linnaeus, Sp. pl. *ed. 2*, 1437. 763.—TYPE (here designated): Woodcut, page 248, vol. *2, Hist. pl.*, 1650, J. Bauhin (lectotype). Linnaeus cited Caspar Bauhin's (1623) Pinax, page 310, Jean Bauhin's (1650) Historia, vol. 2, page 247, and Dodonaeus' (1583) Stirpium historiae pemptades, page 662. Jeffrey (1980) selected the text in Bauhin's Pinax as lectotype, but under the current ICBN (Greuter et al., 1988) that is inaffective. Dodoens (1583) treated *Cucumis oblongus* (the phrase in Linnaeus) and supplied a woodcut. It shows an entire plant including roots, stems, leaves, and ovoid fruits. There are some *C. melo*, s.s., with similar fruits. Caspar Bauhin (1623) treated *Cucumis flexuosus* (the phrase in Linnaeus), and there was no woodcut with it. Jean Bauhin (1650) treated *Cucumis longissimus* starting on page 247 and continuing on page 248. A woodcut was presented on page 248 with separate leaves, a stem apex, a cylindrical fruit, and a seed. Bauhin's (1650) woodcut represented what is now known as *C. flexuosus* Linnaeus. He also cited Dodoens correctly. Since there are no extant specimens (C.E. Jarvis, pers. comm.), Bauhin's woodcut is here selected as lectotype.

Cucumis flexuosus ser. var. *adzhur* Pangalo in Zhukovsky, La Turquie agricole 542. 1933.

Cucumis flexuosus var. *curvato-albus* Pangalo in Zhukovsky, La Turquie agricole 542. 1933.

Cucumis flexuosus var. *curvato-aurantiacus* Pangalo in Zhukovsky, La Turquie agricole 543. 1933.

Cucumis flexuosus var. *curvato-viridis* Pangalo in Zhukovsky, La Turquie agricole 543. 1933.

Cucumis flexuosus var. *recto-albus* Pangalo in Zhukovsky, La Turquie agricole 542. 1933.

Cucumis flexuosus var. *recto-auranticus* Pangalo in Zhukovsky, La Turquie agricole 543. 1933.

Cucumis flexuosus var. *recto-viridis* Pangalo in Zhukovsky, La Turquie agricole 543. 1933.

Cucumis flexuosus var. *reflexus* Seringe in de Candolle, Prodr. 3: 300. 1828.—TYPE: Cultivated specimen (neotype: G-DC, second specimen labeled DE CAND. prodr. 3 p. 300 n. 4 [IDC microfiche 2562!]).

Cucumis flexuosus ser. var. *terra* Pangalo in Zhukovsky, La Turquie agricole 542. 1933.

Cucumis jamaicensis Bertero ex Sprengel, Syst. veg. 3: 46. 1826.

Cucumis jucunda F. Mueller, Trans. Phil. Inst. Victoria 3: 45. 1859.—TYPE: AUSTRALIA. Victoria River, *F. von Mueller s.n.* (neotype K!; neoisotype: K!).

Cucumis laevigatus Chiovenda, Nuov. Giorn. Bot. Ital. n.s. 36: 364. 1929.—TYPE: Ethiopia. Arussi, lungo l'alto Uebi Scebeli, *Basile 151* (Cufodontis, 1965).

Cucumis luzonicus Blanco, Fl. Filip. *ed. 1*, 861; *ed. 2*, 534 (1845). 1837.

Cucumis maculatus Willdenow, Sp. pl. 4: 614. 1805.—TYPE: GERMANY. Plant cultivated at Berlin, Germany, *C.L. Willdenow s.n.* (holotype: B-W [IDC microfiche 7440, specimen number 18049!]).

Cucumis maltensis Seringe ex Steudel, Nomencl. bot. *ed.* 2, 2: 451. 1840. Invalid, published as synonym of *C. melo* Linnaeus.

Cucumis specioid melo Linnaeus emend. Grebenscikov, Kulturpflanze 1: 134. 1953.

Cucumis melo Linnaeus var. *acidulus* Naudin, Ann. Sc. Nat. Ser. 4, 11: 66. 1859.—TYPE (here designated): FRANCE. Cultivated at the Muséum d'Histoire Naturelle, seeds from Pondicheri, India, sent by Jules Lépeire, 1858, *C.V. Naudin s.n.* (lectotype: P; isolectotype: P).

Cucumis melo Linnaeus convar. *adana* (Pangalo) Grebenscikov, Kulturpflanze 1: 134. 1953. Invalid, lacks full and direct reference to place of valid publication of its basionym.

Cucumis melo Linnaeus ser. var. *adana* Pangalo in Zhukovsky, La Turquie Agricole 535. 1933.

Cucumis melo Linnaeus sortentypus *adana* Pangalo, Deutsche Landw. Rundschau 2: 65. 1928.

Cucumis melo Linnaeus subsp. *adana* Pangalo in Vavilov, Theoret. bases pl. breed. 3: 148. 1937.

Cucumis melo Linnaeus var. *adress* Pangalo in Zhukovsky, La Turquie agricole 540. 1933.

Cucumis melo Linnaeus convar. *adzhur* (Pangalo) Grebenscikov, Kulturpflanze 1: 135. 1935.

Cucumis melo Linnaeus convar. *adzhur* (Pangalo) Grebenscikov, Kulturpflanze 1: 135. 1953. Invalid, lacks full and direct reference to place of valid publication of its basionym.

Cucumis melo Linnaeus var. *aegyptiaca* (Sickenberger) Hassib, Cucurbit. Egypt 104. 1938.—BASIONYM: *Cucumis dudaim* Linnaeus *aegyptiaca* Sickenberger.

Cucumis melo Linnaeus subvar. *aegyptiaca* D.S. Fish, Bull. Alexandria Hort. Soc. 6: 125. 1912.

Cucumis melo Linnaeus *aegypticus* Alefeld, Landw. Fl. 204. 1866. Published without indication of rank, so inoperative for priority except for homonymy.

Cucumis melo Linnaeus *aegypticus* C. Harz, Landw. Samenk. 2: 783. 1885. Published without indication of rank, so inoperative for priority except for homonymy.

Cucumis melo Linnaeus var. *aestivales* Filov, Vestn. Skh. Nauki 1: 130. 1960. Illegitimate, later homonym of *C. melo aestivalis* Alefeld since there was no direct reference to it by Filov. Also, a type was not designated.

Cucumis melo Linnaeus *aestivalis* Alefeld, Landw. Fl. 205. 1866. Published without indication of rank, so inoperative for priority except for homonymy.

Cucumis melo Linnaeus var. *aethiopicus* Naudin, Ann. Sc. Nat. Ser.4, 12: 110. 1859.—TYPE (here designated): FRANCE. Cutlivated at the Muséum d'Histoire Naturelle, seeds from Egyptian Sudan sent by Figari-Bey, 1863, *Naudin s.n.* (lectotype: P; isolectotype: P).

Cucumis melo Linnaeus ssp. *agrestis* (Naud.) Pangalo in Zhukovsky, La Turquie agricole 534, 874. 1933.

Cucumis melo Linnaeus subvar. *agrestis* (Naudin) Pangalo, Deutsche Landw. Rundschau 2: 64. 1928.—BASIONYM: *Cucumis melo* Linnaeus var. *agrestis* Naudin.

Cucumis melo Linnaeus var. *agrestis* Naudin, Ann. Sc. Nat. Ser. 4, 11: 73. 1859.—TYPE (here designated): FRANCE. Cultivated at the Muséum d'Histoire Naturelle, seeds from Pondicheri, India, sent by Jules Lépeire, 1859, *Naudin s.n.* (lectotype: P; isolectotype: P).

Cucumis melo Linnaeus var. *albida* Makino, J. Jap. Bot. 5(9): 34. 1928. Illegitimate, later homonym of *C. melo albidus* Alefeld since there was no reference, direct or indirect, to it by Makino.

Cucumis melo Linnaeus *albidus* Alefeld, Landw. Fl. 204. 1866. Published without indication of rank, so inoperative for priority except for homonymy.

Cucumis melo Linnaeus f. *albidus* (Makino) Kitamura, Acta Phytotax. Geobot. 14: 43. 1950. Invalid, the basionym, *C. melo* var. *albida* Makino, illegitimate.

Cucumis melo Linnaeus var. *alboviridis* Pangalo in Zhukovsky, La Turquie agricole 539. 1933.

Cucumis melo Linnaeus f. *albus* Makino, Bot. Magaz. Tokyo 20: 81. 1906.

Cucumis melo Linnaeus *albus-durus* C. Harz, Landw. Samenk. 2: 782. 1885. Published without indication of rank, so inoperative for priority except for homonymy.

Cucumis melo Linnaeus *algericus* C. Harz, Landw. Samenk. 2: 778. 1885. Published without indication of rank, so inoperative for priority except for homonymy.

Cucumis melo Linnaeus convar. *ameri* (Pangalo) Grebenscikov, Kulturpflanze 1: 135. 1953. Invalid, lacks full and direct reference to place of valid publication of its basionym.

Cucumis melo Linnaeus proles *ameri* Pangalo, Melon cultivation in USSR 185. 1933. Nom. nud.

Cucumis melo Linnaeus ser. var. *ameri* Gabaev in Zhukovsky, La Turquie Agicole 540 1933.

Cucumis melo Linnaeus sortentypus *ameri* Pangalo, Deutsche Landw. Rundschau 2: 66. 1928.

Cucumis melo Linnaeus *ananas* Alefeld, Landw. Fl. 208. 1866. Published without indication of rank, so inoperative for priority except for homonymy.

Cucumis melo Linnaeus var. *ananas-í-chair-verte* Hassib, Cucurbit. Egypt 127. 1938. Invalid, lacks description or diagnosis in Latin.

Cucumis melo Linnaeus *ananassa-alba* C. Harz, Landw. Samenk. 2: 782. 1885. Published without indication of rank, so inoperative for priority except for homonymy.

Cucumis melo Linnaeus *ananassa-viridis* C. Harz, Landw. Samenk. 2: 782. 1885. Published without indication of rank, so inoperative for priority except for homonymy.

Cucumis melo Linnaeus var. *anatolicus* Naudin, Ann. Sc. Nat. Ser. 4, 12: 110. 1866.—TYPE (here designated): FRANCE. Cultivated at the Muséum d'Histoire Naturelle, seeds from Africa sent by Figari-Bey, 1864, *Naudin s.n.* (lectotype: P).

Cucumis melo Linnaeus *anglicus* C. Harz, Landw. Samenk. 2: 780. 1885. Published without indication of rank, so inoperative for priority except for homonymy.

Cucumis melo Linnaeus *annae* Alefeld, Landw. Fl. 207. 1866. Published without indication of rank, so inoperative for priority except for homonymy.

Cucumis melo Linnaeus sortentypus *arbaschek* Pangalo, Deutsche Landw. Rundschau 2: 66. 1928.

Cucumis melo Linnaeus *archangelicus* C. Harz, Landw. Samenk. 2: 779. 1885. Published without indication of rank, so inoperative for priority except for homonymy.

Cucumis melo Linnaeus *argenteus* Alfeld, Landw. Fl. 204. 1866. Published without indication of rank, so inoperative for priority except for homonymy.

Cucumis melo Linnaeus *argenteus* C. Harz, Landw. Samenk. 2: 783. 1885. Published without indication of rank, so inoperative for priority except for homonymy.

Cucumis melo Linnaeus var. *arundel* Hassib, Cucurbit. Egypt 127. 1938. Invalid, lacks description or diagnosis in Latin.

Cucumis melo Linnaeus var. *atab* Hassib, Cucurbit. Egypt 132. 1938. Invalid, lacks description or diagnosis in Latin.

Cucumis melo Linnaeus *atossae* Alfeld, Landw. Fl. 203. 1866. Published without indication of rank, so inoperative for priority except for homonymy.

Cucumis melo Linnaeus *aurantiacus* C. Harz, Landw. Samenk. 2: 780. 1885. Published without indication of rank, so inoperative for priority except for homonymy.

Cucumis melo Linnaeus var. *aurantiacus* Pangalo in Zhukovsky, La Turquie agricole 539. 1933. Illegitimate, later homonym of *C. melo aurantiacus* C. Harz since there was no reference, direct or indirect, to it by Pangalo.

Cucumis melo Linnaeus *auratus* (Naudin) C. Harz, Landw. Samenk. 2: 781. 1885. Published without indication of rank, so inoperative for priority except for homonymy.

Cucumis melo Linnaeus var. *autumnales* Filov, Vestn. Skh. Nauki 1: 130. 1960. Invalid, type not designated.

Cucumis melo Linnaeus var. *azmirli* Hassib, Cucurbit. Egypt 129. 1938. Invalid, lacks description or diagnosis in Latin.

Cucumis melo Linnaeus var. *baqubensis* Chakravarty, Iraq Minist. Agric. Techn. Bull. 3: 96. 1977.—TYPE: IRAQ. 8 Oct 1961, *Chakravarty & Nuri 30738* (holotype: BAG).

Cucumis melo Linnaeus subvar. *bassusi* Hassib, Cucurbit. Egypt 109. 1938. Invalid, lacks description or diagnosis in Latin.

Cucumis melo Linnaeus *bathensis* C. Harz, Landw. Samenk. 2: 782. 1885. Published without indication of rank, so inoperative for priority except for homonymy.

Cucumis melo Linnaeus var. *beida* Hassib, Cucurbit. Egypt 93. 1938. Invalid, lacks description or diagnosis in Latin.

Cucumis melo Linnaeus var. *bitariana* (Sickenberger) Hassib, Cucurbit. Egypt 99. 1938.—BASIONYM: *Cucumis dudaim* Linnaeus *bitariana* Sickenberger.

Cucumis melo Linnaeus var. *bos-valdy* Suzuki, Foundation of Hothouse Melon Culture 37. 1970. Invalid, published without indication of rank.

Cucumis melo Linnaeus *brasiliensis* Alfeld, Landw. Fl. 207. 1866. Published without indication of rank, so inoperative for priority except for homonymy.

Cucumis melo Linnaeus var. *buharici* (Pangalo) Filov, Vestn. Skh. Nauki 1: 130. 1960. Invalid, basionym invalid and type not designated.

Cucumis melo Linnaeus var. *bullock* Hassib, Cucurbit. Egypt 127. 1938. Invalid, lacks description or diagnosis in Latin.

Cucumis melo Linnaeus var. *burrell's-gem* Hassib, Cucurbit. Egypt 127. 1938. Invalid, lacks description or diagnosis in Latin.

Cucumis melo Linnaeus *cabulensis* C. Harz, Landw. Samenk. 2: 781. 1885. Published without indication of rank, so inoperative for priority except for homonymy.

Cucumis melo Linnaeus convar. *cantalupa* (Pangalo) Grebenscikov, Kulturpflanze 1: 135. 1953. Invalid, lacks full and direct reference to place of valid publication of its basionym.

Cucumis melo Linnaeus proles *cantalupa* Pangalo, nom. nud., Melon cultivation in USSR 185. 1933.

Cucumis melo Linnaeus ser. var. *cantalupa* Pangalo ex Gabaiev in Zhukovsky, La Turquie Agicole 537. 1933.

Cucumis melo Linnaeus var. *cantalupensis* Naudin, Ann. Sc. Nat. Ser. 4, 11: 47. 1859.

Cucumis melo Linnaeus var.-gr. *cantalupensis* (Naudin) Alefeld, Landw. Fl. 204. 1866.—BASIONYM: *Cucumis melo* Linnaeus var. *cantalupensis* Naudin.

Cucumis melo Linnaeus var. *cantalupo* Seringe in de Candolle, Prodr. 3: 300. 1828.

Cucumis melo Linnaeus var. *cantonianus* Naudin, Ann. Sc. Nat. Ser. 4, 16: 161. 1862.—TYPE (here designated): FRANCE. Cultivated in the Muséum d'Histoire Naturelle, seeds from Canton, China, sent by Fontanier, 1861, *Naudin s.n.* (lectotype: P).

Cucumis melo Linnaeus *carmelitarum* (Naudin) C. Harz, Landw. Samenk. 2: 779. 1885. Published without indication of rank, so inoperative for priority except for homonymy.

Cucumis melo Linnaeus *carmelitarum* Alefeld, Landw. Fl. 208. 1866. Published without indication of rank, so inoperative for priority except for homonymy.

Cucumis melo Linnaeus *caroliniae* (Naudin) C. Harz, Landw. Samenk. 2: 781. 1885. Published without indication of rank, so inoperative for priority except for homonymy.

Cucumis melo Linnaeus *carthagenens* Alefeld, Landw. Fl. 205. 1866. Published without indication of rank, so inoperative for priority except for homonymy.

Cucumis melo Linnaeus ser. var. *casaba* Pangalo in Zhukovsky, La Turquie Agricole 535. 1933.

Cucumis melo Linnaeus *caseiformis* Alefeld, Landw. Fl. 206. 1866. Published without indication of rank, so inoperative for priority except for homonymy.

Cucumis melo Linnaeus convar. *cassaba* (Pangalo) Grebenscikov, Kulturpflanze 1: 134. 1953. Invalid, lacks full and direct reference to place of valid publication of its basionym.

Cucumis melo Linnaeus proles *cassaba* Pangalo, nom. nud., Melon cultivation in USSR 185. 1933.

Cucumis melo Linnaeus sortentypus *cassaba* Pangalo, Deutsche Landw. Rundschau 2: 65. 1928.

Cucumis melo Linnaeus *cassabar* Alefeld, Landw. Fl. 207. 1866. Published without indication of rank, so inoperative for priority except for homonymy.

Cucumis melo Linnaeus *chammam-roumi* C. Harz, Landw. Samenk. 2: 785. 1885. Published without indication of rank, so inoperative for priority except for homonymy.

Cucumis melo Linnaeus sortentypus *chanaljac* Pangalo, Cucumbers 28 [in Russian]. 1928. Invalid, epithet published in Russian.

Cucumis melo Linnaeus convar. *chandalak* (Pangalo) Grebenscikov, Kulturpflanze 1: 135. 1953. Invalid, lacks full and direct reference to place of valid publication of its basionym.

Cucumis melo Linnaeus ser. var. *chandaliak* Gabaiev in Zhukovsky, La Turquie Agicole 540 1933.

Cucumis melo Linnaeus proles *chandaljak* Pangalo, nom. nud., Melon cultivation in USSR 185. 1933.

Cucumis melo Linnaeus var. *chandlkak* (Pangalo) Filov, Vestn. Skh. Nauki 1: 130. 1960.

Cucumis melo Linnaeus var. *chandulak* (Pangalo) Grebenscikov in Suzuki, Foundation of Hothouse Melon Culture, p. 5. 1970. Orthographic variant of *Cucumis melo* Linnaeus proles *chandlajak* Pangalo.

Cucumis melo Linnaeus *chate* (Linnaeus) Harz, Landw. Samenk. 2: 785. 1885. Published without indication of rank, so inoperative for priority except for homonymy.

Cucumis melo Linnaeus subsp. *chate* (Hasselquist) Hassib, Cucurbit. Egypt 133. 1938.

Cucumis melo Linnaeus var. *chate* (Hasselquist) Sageret, Mém. Agric. Soc. Roy. Centr. Aric. 58: 488. 1825.

Cucumis melo Linnaeus *chilensis* Alefeld, Landw. Fl. 207. 1866. Published without indication of rank, so inoperative for priority except for homonymy.

Cucumis melo Linnaeus conv. *chinensis* (Pangalo) Grebenscikov, Kulturpflanze 1: 136. 1953. Invalid, the basionym, *C. chinensis* Pangalo, illegitimate.

Cucumis melo Linnaeus convar. *chinensis* (Pangalo) Grebenscikov, Kulturpflanze 1: 136. 1953. Invalid, lacks full and direct reference to place of valid publication of its basionym.

Cucumis melo Linnaeus ssp. *chinensis* (Pangalo) Filov, Vestn. Skh. Nauki 1: 131. 1960. Invalid, the basionym, *Cucumis chinensis* Pangalo, illegitimate.

Cucumis melo Linnaeus subsp. *chinensis* (Pangalo) Pangalo, Trudy Prikl. Bot. 23(3): 231, 250. 1930.—BASIONYM: *Cucumis chinensis* Pangalo. Invalid, the basionym is illegitimate.

Cucumis melo Linnaeus var. *chinensis* (Naudin) Pangalo, Deutsche Landw. Rundschau 2: 64. 1928. Invalid, the basionym, *C. chinensis* Pangalo, illegitimate.

Cucumis melo Linnaeus var. *chito* (C. Morren) Naudin, Ann. Sc. Nat. Ser. 4, 11: 67. 1859.

Cucumis melo Linnaeus *chloris* Alefeld, Landw. Fl. 203. 1866. Published without indication of rank, so inoperative for priority except for homonymy.

Cucumis melo Linnaeus *chlorosarcinus* C. Harz, Landw. Samenk. 2: 784. 1885. Published without indication of rank, so inoperative for priority except for homonymy.

Cucumis melo Linnaeus *christianiensis* C. Harz, Landw. Samenk. 2: 781. 1885. Published without indication of rank, so inoperative for priority except for homonymy.

Cucumis melo Linnaeus *chrysopetrus* Alefeld, Landw. Fl. 205. 1866. Published without indication of rank, so inoperative for priority except for homonymy.

Cucumis melo Linnaeus *citrinus* Alefeld, Landw. Fl. 204. 1866. Published without indication of rank, so inoperative for priority except for homonymy.

Cucumis melo Linnaeus *citrodorus* Alefeld, Landw. Fl. 206. 1866. Published without indication of rank, so inoperative for priority except for homonymy.

Cucumis melo Linnaeus *communis* C. Harz, Landw. Samenk. 2: 778. 1885. Published without indication of rank, so inoperative for priority except for homonymy.

Cucumis melo Linnaeus convar. *conomon* (Thunberg) Grebenscikov, Kulturpflanze 1: 135. 1953. Invalid, lacks full and direct reference to basionym publication.

Cucumis melo Linnaeus subspecoid *conomon* (Thunberg) Grebenscikov, Kulturpflanze 1: 135. 1935. Invalid, lacks full and direct reference to basionym publication.

Cucumis melo Linnaeus var. *conomon* (Thunberg) Makino, Bot. Magaz. Tokyo 16: 16. 1902.

Cucumis melo Linnaeus *coronata* Alefeld, Landw. Fl. 201. 1866. Published without indication of rank, so inoperative for priority except for homonymy.

Cucumis melo Linnaeus *coronatus* C. Harz, Landw. Samenk. 2: 784. 1885.—Invalid, published without indication of rank.

Cucumis melo Linnaeus var. *cossonianus* Naudin, Ann. Sc. Nat. Ser. 4, 16: 155. 1862.—TYPE (here designated): FRANCE. Cultivated at the Muséum d'Histoire Naturelle, seeds sent by Cosson, 1860, *Naudin s.n.* (lectotype: P; isolectotype: P).

Cucumis melo Linnaeus *cossonii* C. Harz, Landw. Samenk. 2: 785. 1885. Published without indication of rank, so inoperative for priority except for homonymy.

Cucumis melo Linnaeus *coulommierensis* C. Harz, Landw. Samenk. 2: 778. 1885. Published without indication of rank, so inoperative for priority except for homonymy.

Cucumis melo Linnaeus var. *cucurbitaceus* Chakravarty, Iraq Minist. Agric. Tech. Bull. 3: 91. 1977.—TYPE: IRAQ. Samarra, 4 Nov 1961, *Chakravarty & Nuri 30739* (holotype, BAG).

Cucumis melo Linnaeus var. *cultus* W.S. Kurz, Journ. Asiat. Soc. Beng. 46(2): 103. 1877.

Cucumis melo Linnaeus *cultus* (W.S. Kurz) Pangalo in Zhukovsky, La Turquie agricole 534. 1933. Published without indication of rank, so inoperative for priority except for homonymy.

Cucumis melo Linnaeus ssp. *cultus* (W.S. Kurz) Pangalo in Zhukovsky, La Turquie agricole 534. 1933.

Cucumis melo Linnaeus subvar. *cultus* (W.S. Kurz) Pangalo, Deutsche Landw. Rundschau 2: 64. 1928.—BASIONYM: *Cucumis melo* Linnaeus var. *cultus* Kurtz.

Cucumis melo Linnaeus *cypricus* (Naudin) C. Harz, Landw. Samenk. 2: 781. 1885. Published without indication of rank, so inoperative for priority except for homonymy.

Cucumis melo Linnaeus *dampsa* Alefeld, Landw. Fl. 203. 1866. Published without indication of rank, so inoperative for priority except for homonymy.

Cucumis melo Linnaeus *daree* Alefeld, Landw. Fl. 203. 1866. Published without indication of rank, so inoperative for priority except for homonymy.

Cucumis melo Linnaeus *deliciosus* Alefeld, Landw. Fl. 200. 1866. Published without indication of rank, so inoperative for priority except for homonymy.

Cucumis melo Linnaeus *deltae* Sickenberger, Mem. Inst. Eqypt. 4(2): 234. 1901. Published without indication of rank, so inoperative for priority except for homonymy.

Cucumis melo Linnaeus f. *depressus* Chakravarty, Iraq. Minist. Agric. Tech. Bull. 3: 92. 1977.—TYPE: IRAQ. Kochek, 8 Oct 1961, *Chakravarty & Nuri 30740* (holotype: BAG).

Cucumis melo Linnaeus *dinodaim* Alefeld, Landw. Fl. 200. 1866. Published without indication of rank, so inoperative for priority except for homonymy.

Cucumis melo Linnaeus f. *dissectifolius* Pangalo, Trudy Prikl. Bot. 23(3): 244. 1930.

Cucumis melo Linnaeus *ducalis* Alefeld, Landw. Fl. 207. 1866. Published without indication of rank, so inoperative for priority except for homonymy.

Cucumis melo Linnaeus subspecoid *dudaim* (Linnaeus) Grebenscikov, Kulturpflanze 1: 134. 1953.—BASIONYM: *Cucumis dudaim* Linnaeus. Invalid, lacks full and direct reference to place of valid publication of its basionym.

Cucumis melo Linnaeus var. *dudaim* (Linnaeus) Naudin, Ann. Sc. Nat. Ser. 4, 11: 69. 1859.—BASIONYM: *Cucumis dudaim* Linnaeus.

Cucumis melo Linnaeus var.-gr. *dudaim* (linnaeus) Alefeld, Landw. Fl. 199. 1866.—BASIONYM: *Cucumis dudaim* Linnaeus.

Cucumis melo Linnaeus f. *dulcis* Pangalo, Trudy Prikl. Bot. 23(3): 244. 1930.

Cucumis melo Linnaeus var. *dumeri* Hassib, Cucurbit. Egypt 128. 1938. Invalid, lacks description or diagnosis in Latin.

Cucumis melo Linnaeus *durans* Alefeld, Landw. Fl. 202. 1866. Published without indication of rank, so inoperative for priority except for homonymy.

Cucumis melo Linnaeus var. *duripulposus* Filov, Vestn. Skh. Nauki 1: 130. 1960. Invalid, type not designated.

Cucumis melo Linnaeus *durrieni* Alefeld, Landw. Fl. 199. 1866. Published without indication of rank, so inoperative for priority except for homonymy.

Cucumis melo Linnaeus sortentypus *dutma* Pangalo, Deutsche Landw. Rundschau 2: 65. 1928.

Cucumis melo Linnaeus *ecostatus* Alefeld, Landw. Fl. 208. 1866. Published without indication of rank, so inoperative for priority except for homonymy.

Cucumis melo Linnaeus var. *elongata* (Sickenberger) Hassib, Cucurbit. Egypt 100, *f. 42.* 1938.—BASIONYM: *Cucumis dudaim* Linnaeus *elongata* Sickenberger.

Cucumis melo Linnaeus var.-gr. *elongatus* Alefeld, Landw. Fl. 201. 1866.

Cucumis melo Linnaeus subvar. *elongatus* D.S. Fish, Bull. Alexandria Hort. Soc. 6: 126. 1912.

Cucumis melo Linnaeus *endaurantius* Alefeld, Landw. Flora 208. 1866. Published without indication of rank, so inoperative for priority except for homonymy.

Cucumis melo Linnaeus var. *erythraeus* Naudin, Ann. Sc. Nat. Ser. 4, 11: 73. 1859.—TYPE (here designated): FRANCE. Cultivated at the Muséum d'Histoire Naturelle, seeds from England sent by Lindley, 1858, *C.V. Naudin s.n.* (lectotype: P; isolectotype: P).

Cucumis melo Linnaeus *erythrosarcinus* C. Harz, Landw. Samenk. 2: 784. 1885. Published without indication of rank, so inoperative for priority except for homonymy.

Cucumis melo Linnaeus *erythrosarx* Alefeld, Landw. Fl. 205. 1866. Published without indication of rank, so inoperative for priority except for homonymy.

Cucumis melo Linnaeus subsp. *europaeus* Filov, Vestn. Skh. Nauki 1: 130. 1960. Invalid, type not designated.

Cucumis melo Linnaeus *facqous* C. Harz, Landw. Samenk. 2: 785. 1885. Published without indication of rank, so inoperative for priority except for homonymy.

Cucumis melo Linnaeus *faginus* Alefeld, Landw. Fl. 208. 1866. Published without indication of rank, so inoperative for priority except for homonymy.

Cucumis melo Linnaeus *fairi* Alefeld, Landw. Fl. 209. 1866. Published without indication of rank, so inoperative for priority except for homonymy.

Cucumis melo Linnaeus var. *firany* Hassib, Cucurbit. Egypt 91, *f. 37.* 1938. Invalid, lacks description or diagnosis in Latin.

Cucumis melo Linnaeus var. *flava* Makino, Journ. Jap. Bot. 5: 32. 1928.

Cucumis melo Linnaeus f. *flavus* (Makino) Kitamura, Acta Phytotax. Geobot. 14: 43. 1950.

Cucumis melo Linnaeus var. *flavus* Pangalo in Zhukovsky, La Turquie agricole 539. 1933. Illegitimate, later homonym.

Cucumis melo Linnaeus convar. *flexuosus* (Linnaeus) Grebenscikov, Kulturpflanze 1: 135. 1935.

Cucumis melo Linnaeus subsp. *flexuosus* (Linnaeus) Filov, Vestn. Skh. Nauki 1: 131. 1960.—BASIONYM: *Cucumis flexuosus* Linnaeus. Invalid, lacks full and direct reference to place of valid publication of its basionym.

Cucumis melo Linnaeus subsp. *flexuosus* (Linnaeus) Pangalo, Trudy Prikl. Bot. 23(3): 231, 250. 1930.—BASIONYM: *Cucumis flexuosus* Linnaeus.

Cucumis melo Linnaeus subspecoid *flexuosus* (Linnaeus) Grebenscikov, Kulturpflanze 1: 135. 1953.—BASIONYM: *Cucumis flexuosus* Linnaeus. Invalid, lacks full and direct reference to place of valid publication of its basionym.

Cucumis melo Linnaeus var. *flexuosus* (Linnaeus) Naudin, Ann. Sc. Nat. Ser. 4, 11: 63. 1859.—BASIONYM: *Cucumis flexuosus* Linnaeus.

Cucumis melo Linnaeus var. *fraiduni* Chakravarty, Iraq Minist. Agric. Tech. Bull. 3: 94. 1977.—TYPE: IRAQ. Samarra, *Chakravarty & Nuri 30724* (holotype: BAG); *30725* (paratype: BAG); *30726* (paratype: BAG).

Cucumis melo Linnaeus *friesicus* Alefeld, Landw. Fl. 206. 1866. Published without indication of rank, so inoperative for priority except for homonymy.

Cucumis melo Linnaeus var. *fucharici* Suzuki, Foundation of Hothouse Melon Culture 5. 1970. Nom. nud.

Cucumis melo Linnaeus *gallicus* Alefeld, Landw. Fl. 205. 1866. Published without indication of rank, so inoperative for priority except for homonymy.

Cucumis melo Linnaeus *garmack* C. Harz, Landw. Samenk. 2: 784. 1885.—Invalid, published without indication of rank.

Cucumis melo Linnaeus *gehoon* Alefeld, Landw. Fl. 205. 1866. Published without indication of rank, so inoperative for priority except for homonymy.

Cucumis melo Linnaeus var. *gem-rocky* Hassib, Cucurbit. Egypt 127. 1938. Invalid, lacks description or diagnosis in Latin.

Cucumis melo Linnaeus *georgii* C. Harz, Landw. Samenk. 2: 781. 1885. Published without indication of rank, so inoperative for priority except for homonymy.

Cucumis melo Linnaeus *gerger* Alefeld, Landw. Fl. 203. 1866. Published without indication of rank, so inoperative for priority except for homonymy.

Cucumis melo Linnaeus *germek* Alefeld, Landw. Fl. 203. 1866. Published without indication of rank, so inoperative for priority except for homonymy.

Cucumis melo Linnaeus subvar. *geyeidy* Hassib, Cucurbit. Egypt 114. 1938. Invalid, lacks description or diagnosis in Latin.

Cucumis melo Linnaeus *giganteus* Alefeld, Landw. Fl. 208. 1866. Published without indication of rank, so inoperative for priority except for homonymy.

Cucumis melo Linnaeus f. *ginmakua* Kitamura, Acta Phytotax. Geobot. 14: 43. 1950.

Cucumis melo Linnaeus var. *globosus* Sickenberger, Mem. Inst. Egypt. 4(2): 234. 1901. Published without indication of rank, so inoperative for priority except for homonymy.

Cucumis melo Linnaeus var. *golobcresdael* Suzuki, Foundation of Hothouse Melon Culture 7. 1970. Nom. nud.

Cucumis melo Linnaeus var. *goorgab* Alefeld, Landw. Fl. 203. 1866. Published without indication of rank, so inoperative for priority except for homonymy.

Cucumis melo Linnaeus var. *gourbek* Filov, Vestn. Skh. Nauki 1: 131. 1960. Invalid, type not designated.

Cucumis melo Linnaeus var. *gracilior* Pangalo, Deutsche Landw. Rundschau 2: 62. 1928. Published without indication of rank, so inoperative for priority except for homonymy.

Cucumis melo Linnaeus var. *gracilior* Pangalo, Trudy Prikl. Bot. 23(3): 244. 1930.

Cucumis melo Linnaeus *graciliores* Pangalo, Russ. Bot. Congress, Leningrad, Jan 1928, 301. 1928. Published without indication of rank, so inoperative for priority except for homonymy.

Cucumis melo Linnaeus f. *grandifolius* Pangalo, Trudy Prikl. Bot. 23(3): 244. 1930.

Cucumis melo Linnaeus var. *hafednafse* Chakravarty, Iraq Minist. Agric. Tech. Bull. 3: 90. 1977.—TYPES: IRAQ. Abu Ghraib, Samarra, *Chakravarty & Hazim 30714* (holotype: BAG); 4 Nov 1961, *30716* (paratype: BAG); Dehni Abu Ghraib, 2 Aug 1961, *Chakravarty & Nuri 30713* (paratype: BAG).

Cucumis melo Linnaeus f. *hamikua* Kitamura, Acta Phytotax. Geobot. 14: 43. 1950.

Cucumis melo Linnaeus *hasanbey* Pangalo in Zhukovsky, La Turquie agricole 539, 876. 1933.

Cucumis melo Linnaeus *hermonthicus* Sickenberger, Mem. Inst. Egypt. 4(2): 234. 1901. Published without indication of rank, so inoperative for priority except for homonymy.

Cucumis melo Linnaeus var. *hibernus* Filov, Vestn. Skh. Nauki 1: 130. 1960. Invalid, type not designated.

Cucumis melo Linnaeus var. *hiemalis* Filov, Vestn. Skh. Nauki 1: 130. 1960. Invalid, type not designated.

Cucumis melo Linnaeus var. *hime* Makino in Linuma, Somoku-Dzusetsu 4(20): 1308. 1912.—TYPE (here designated): *P. 40, Somoku-Dzusetsu.* 1912 (lectotype).

Cucumis melo Linnaeus *hollandicus* C. Harz, Landw. Samenk. 2: 778. 1885. Published without indication of rank, so inoperative for priority except for homonymy.

Cucumis melo Linnaeus var. *honey-dew* Hassib, Cucurbit. Egypt 130. 1938. Invalid, lacks description or diagnosis in Latin.

Cucumis melo Linnaeus *hooseini* Alefeld, Landwirthschaftliche Flora 203. 1866. Published without indication of rank, so inoperative for priority except for homonymy.

Cucumis melo Linnaeus *hunteri* C. Harz, Landw. Samenk. 2: 781. 1885. Published without indication of rank, so inoperative for priority except for homonymy.

Cucumis melo Linnaeus *hybridus-persicus* C. Harz, Landw. Samenk. 2: 782. 1885. Published without indication of rank, so inoperative for priority except for homonymy.

Cucumis melo Linnaeus var. *indica* Suzuki, Foundation of Hothouse Melon Culture 37. 1970.—Nom. nud.

110   Appendixes

Cucumis melo Linnaeus *indicus* Alefeld, Landw. Fl. 207. 1866.—Invalid, published without indication of rank.
Cucumis melo Linnaeus *indo-persicus* C. Harz, Landw. Samenk. 2: 784. 1885.—Invalid, published without indication of rank.
Cucumis melo Linnaeus var. *inodorus* H. Jacquin, Monogr. melon 173. 1832.
Cucumis melo Linnaeus *inodorus* C. Harz, Landw. Samenk. 2: 783. 1885. Illegitemate, later homonym.
Cucumis melo Linnaeus *insipidus* Alefeld, Landw. Fl. 200. 1866. Published without indication of rank, so inoperative for priority except for homonymy.
Cucumis melo Linnaeus *insipidus* Naudin ex C. Harz, Landw. Samenk. 2: 785. 1885. Illegitemate, later homonym.
Cucumis melo Linnaeus *insularis* Alefeld, Landw. Fl. 202. 1866. Published without indication of rank, so inoperative for priority except for homonymy.
Cucumis melo Linnaeus *ispahanensis* Alefeld, Landw. Fl. 203. 1866. Published without indication of rank, so inoperative for priority except for homonymy.
Cucumis melo Linnaeus *italicus* Alefeld, Landw. Fl. 206. 1866. Published without indication of rank, so inoperative for priority except for homonymy.
Cucumis melo Linnaeus *jalibie* C. Harz, Landw. Samenk. 2: 754. 1885. Published without indication of rank, so inoperative for priority except for homonymy.
Cucumis melo Linnaeus *japonicus* C. Harz, Landw. Samenk. 2: 781. 1885. Published without indication of rank, so inoperative for priority except for homonymy.
Cucumis melo Linnaeus *jileyi* C. Harz, Landw. Samenk. 2: 784. 1885.—Invalid, published without indication of rank.
Cucumis melo Linnaeus *keiseng* Alefeld, Landw. Fl. 203. 1866. Published without indication of rank, so inoperative for priority except for homonymy.
Cucumis melo Linnaeus var. *khadra* Hassib, Cucurbit. Egypt 94, *f. 41*. 1938. Invalid, lacks description or diagnosis in Latin.
Cucumis melo Linnaeus f. *kikumelon* Kitamura, Acta Phytotax. Geobot. 14: 43. 1950.
Cucumis melo Linnaeus var. *kirukensis* Chakravarty, Iraq Minist. Agric. Tech. Bull. 3: 90. 1977.—TYPES: IRAQ. Tuk, 12 Oct 1961, *Chakravarty & Nuri 30743* (holotype: BAG); Sangar Chebier, 29 Oct 1961, *Chakravarty & Khatib 30744* (paratype: BAG); Eskikelek, 8 Oct 1961, *Chakravarty & Nuri 30742* (paratype: BAG).
Cucumis melo Linnaeus *kurschaing* Alefeld, Landw. Fl. 203. 1866. Published without indication of rank, so inoperative for priority except for homonymy.
Cucumis melo Linnaeus *leeanus* Alefeld, Landw. Fl. 206. 1866. Published without indication of rank, so inoperative for priority except for homonymy.
Cucumis melo Linnaeus *leptodictyus* C. Harz, Landw. Samenk. 2: 785. 1885.—Invalid, published without indication of rank.
Cucumis melo Linnaeus *leucadicus* Alefeld, Landw. Fl. 206. 1866. Published without indication of rank, so inoperative for priority except for homonymy.
Cucumis melo Linnaeus *leucodictyon* Alefeld, Landw. Fl. 200. 1866. Published without indication of rank, so inoperative for priority except for homonymy.
Cucumis melo Linnaeus *leucosarcinus* C. Harz, Landw. Samenk. 2: 784. 1885. Published without indication of rank, so inoperative for priority except for homonymy.
Cucumis melo Linnaeus *leucospermus* Alefeld, Landw. Fl. 206. 1866. Published without indication of rank, so inoperative for priority except for homonymy.
Cucumis melo Linnaeus *liliinus* Alefeld, Landw. Fl. 202. 1866. Published without indication of rank, so inoperative for priority except for homonymy.
Cucumis melo Linnaeus f. *longus* Chakravarty, Iraq Minist. Agric. Tech. Bull. 3: 90. 1977.—TYPE: IRAQ. Samarra, 24 Aug 1961, *Chakravarty & Nuri 30715* (holotype: BAG).
Cucumis melo Linnaeus var. *longus* Chakravarty, Iraq Minist. Agric. Tech. Bull. 3: 94. 1977.—TYPE: IRAQ. Samarra, 24 Aug 1961, *Chakravarty & Nuri 30729* (holotype: BAG).
Cucumis melo Linnaeus var. *macro-castanus* Pangalo in Zhukovsky, La Turquie agricole 536. 1933.
Cucumis melo Linnaeus var. *macro-leucus* Pangalo in Zhukovsky, La Turquie agricole 536. 1933.
Cucumis melo Linnaeus var. *macro-pyrochrus* Pangalo in Zhukovsky, La Turquie agricole 536. 1933.
Cucumis melo Linnaeus var. *maculatus* Naudin, Ann. Sc. Nat. Ser. 4, 12: 110. 1859.
Cucumis melo Linnaeus f. *major* Chakravarty, Iraq Minist. Agric. Tech. Bull. 3: 91. 1977.—TYPE: IRAQ. South of Mosul, river side, *Chakravarty 19651* (holotype: BAG).
Cucumis melo Linnaeus var. *major* Chakravarty, Iraq Minist. Agric. Tech. Bull. 3: 93. 1977.—Types: Iraq, Khan Beni Saad, 11 Oct 1961, *Chakravarty & Nuri 30719* (holotype: BAG); Sangar Chebier, 29 Oct 1961, *Chakravarty & Khatib 30720* (paratype: BAG); Market-Mosul, 9 Oct 1961, *Chakravarty & Nuri 30732* (paratype: BAG).
Cucumis melo Linnaeus var. *makua* Kitamura ex Makino, Acta Phytotax. Geobot. 14: 43. 1950. Orthographic variant of *Cucumis melo* L. var. *makuwa* Makino.
Cucumis melo Linnaeus var. *makuwa* Makino, J. Jap. Bot. 5(8): 32. 1928.
Cucumis melo Linnaeus var. *maltensis* Seringe in de Candolle, Prodr. 3: 300. 1828.
Cucumis melo Linnaeus var. *mansouri* Hassib, Cucurbit. Egypt 119. 1938. Invalid, lacks description or diagnosis in Latin.
Cucumis melo Linnaeus *maurusius* C. Harz, Landw. Samenk. 2: 780. 1885. Published without indication of rank, so inoperative for priority except for homonymy.

Cucumis melo Linnaeus var. *mehanawy* Hassib, Cucurbit. Egypt 132. 1938. Invalid, lacks description or diagnosis in Latin.

Cucumis melo Linnaeus var.-gr. *melitensis* Seringe ex Alefeld, Landw. Fl. 201. 1866. Orthographic variant of *Cucumis melo* Linnaeus var. *maltensis* Seringe in de Candolle.

Cucumis melo Linnaeus *melitensis* C. Harz, Landw. Samenk. 2: 784. 1885. Published without indication of rank, so inoperative for priority except for homonymy.

Cucumis melo Linnaeus *melitensis* Seringe ex Fujishita & Oda, Results of the Kyoto University Scientific Expedition to Karakorum and Hindukushi 1: 254. 1965. Orthographic variant of *Cucumis melo* Linnaeus var. *maltensis* Seringe in de Candolle.

Cucumis melo Linnaeus var.-gr. *microcarpus* Alefeld, Landw. Fl. 198. 1866.

Cucumis melo Linnaeus *microcarpus* C. Harz, Landw. Samenk. 2: 782. 1885. Published without indication of rank, so inoperative for priority except for later homonymy.

Cucumis melo Linnaeus subsp. *microcarpus* (Alefeld) Pangalo, Trudy Prikl. Bot. 23(3): 231, 250. 1930.—BASIONYM: *Cucumis melo* Linnaeus var.-gr. *microcarpus* Alefeld.

Cucumis melo Linnaeus var. *microcarpus* (Alefeld) Pangalo, Deutsche Landw. Rundschau 2: 63. 1928.—BASIONYM: *Cucumis melo* Linnaeus var.-gr. *microcarpus* Alfeld.

Cucumis melo Linnaeus var. *micro-castanus* Pangalo in Zhukovsky, La Turquie agricole 536. 1933.

Cucumis melo Linnaeus var. *micro-leucus* Pangalo in Zhukovsky, La Turquie agricole 537. 1933.

Cucumis melo Linnaeus *micropersicus* C. Harz, Landw. Samenk. 2: 784. 1885. Published without indication of rank, so inoperative for priority except for homonymy.

Cucumis melo Linnaeus var. *micro-pyrochrus* Pangalo in Zhukovsky, La Turquie agricole 536. 1933.

Cucumis melo Linnaeus *microspermus* Alefeld, Landw. Fl. 208. 1866. Published without indication of rank, so inoperative for priority except for homonymy.

Cucumis melo Linnaeus *microspermus* C. Harz, Landw. Samenk. 2: 782. 1885. Illegitemate, later homonym.

Cucumis melo Linnaeus var. *microspermus* Nakai ex Kitamura, Acta Phytotax. Geobot. 14: 43. 1950. Illegitimate, later homonym of *C. melo microspermus* Alefeld since there was no reference, direct or indirect, to it by Kitamura.

Cucumis melo Linnaeus var. *minimus* Chakravarty, Iraq Minist. Agric. Tech. Bull. 3: 96. 1977.—TYPE: IRAQ. Suati Aksu (Tuj), 11 Oct 1961, *Chatravarty & Nuri 30724* (holotype: BAG).

Cucumis melo Linnaeus f. *minor* Chakravarty, Iraq Minist. Agric. Tech. Bull. 3: 90. 1977.—TYPE: IRAQ. Samarra, 24 Aug 1961, *Chakravarty & Nuri 30717* (holotype: BAG).

Cucumis melo Linnaeus var. *minutissimus* Naudin, Ann. Sci. Nat., ser.4, 16: 156. 1862.—TYPE (here designated): FRANCE. Cultivated at the Muséum d'Histoire Naturelle, 1863, *Naudin s.n.* (lectotype: P).

Cucumis melo Linnaeus *mogolensis* C. Harz, Landw. Samenk. 2: 778. 1885. Published without indication of rank, so inoperative for priority except for homonymy.

Cucumis melo Linnaeus var. *momordica* (Roxburgh) Duthie & Fuller, Field Garden Crops 2: 50, *t. 49.* 1883.—BASIONYM: *Cucumis momordica* Roxburgh.

Cucumis melo Linnaeus var. *monoclinus* (Pangalo) Filov, Vestn. Skh. Nauki 1: 131. 1960.—BASIONYM: *Melo monoclinus* Pangalo.

Cucumis melo Linnaeus *monspeliacus* C. Harz, Landw. Samenk. 2: 782. 1885. Published without indication of rank, so inoperative for priority except for homonymy.

Cucumis melo Linnaeus *montaguanus* Alefeld, Landw. Fl. 204. 1866. Published without indication of rank, so inoperative for priority except for homonymy.

Cucumis melo Linnaeus *moschatus* C. Harz, Landw. Samenk. 2: 779. 1885. Published without indication of rank, so inoperative for priority except for homonymy.

Cucumis melo Linnaeus *naudini* Alefeld, Landw. Fl. 200. 1866. Published without indication of rank, so inoperative for priority except for homonymy.

Cucumis melo Linnaeus f. X *neomakua* Kitamura, Acta Phytotax. Geobot. 14: 43. 1950. Parents: *Cucumis melo* L. var. *typica* X var. *makuwa* Makino.

Cucumis melo Linnaeus *nigrescens* Alefeld, Landw. Fl. 207. 1866. Published without indication of rank, so inoperative for priority except for homonymy.

Cucumis melo Linnaeus *nucinus* Alefeld, Landw. Fl. 207. 1866. Published without indication of rank, so inoperative for priority except for homonymy.

Cucumis melo Linnaeus *nuksewan* Alefeld, Landw. Fl. 203. 1866. Published without indication of rank, so inoperative for priority except for homonymy.

Cucumis melo Linnaeus *oblongus* Sickenberger, Mem. Inst. Egypt. 4(2): 234. 1901. Published without indication of rank, so inoperative for priority except for homonymy.

Cucumis melo Linnaeus var. *oblongus* Chakravarty, Iraq Minist. Agric. Tech. Bull. 3: 89. 1977.—TYPE: IRAQ. Samarra, 24 Aug 1961, *Chakravarty & Nuri 30718* (holotype: BAG). Illegitimate, later homonym.

Cucumis melo Linnaeus *odessana* Alefeld, Landw. Fl. 204. 1866. Published without indication of rank, so inoperative for priority except for homonymy.

Cucumis melo Linnaeus *odoratissimus* Alefeld, Landw. Fl. 200. 1866. Published without indication of rank, so inoperative for priority except for homonymy.

Cucumis melo Linnaeus *olorinus* Alefeld, Landw. Fl. 205. 1866. Published without indication of rank, so inoperative for priority except for homonymy.

Cucumis melo Linnaeus ssp. *orientale* Sageret ex Filov, Vestn. Skh. Nauki 131. 1960. Invalid, type not designated.
Cucumis melo Linnaeus *orientalis* Alefeld, Landw. Fl. 206. 1866. Published without indication of rank, so inoperative for priority except for homonymy.
Cucumis melo Linnaeus var. *ovatus* Chakravarty, Iraq Minist. Agric. Tech. Bull. 3: 94. 1977.—TYPE: IRAQ. Samarra, 24 Sep 1961, *Chakravarty & Nuri 30730* (holotype: BAG).
Cucumis melo Linnaeus *pachyderma* Alefeld, Landw. Fl. 205. 1866. Published without indication of rank, so inoperative for priority except for homonymy.
Cucumis melo Linnaeus *paplis* Alefeld, Landw. Fl. 209. 1866. Published without indication of rank, so inoperative for priority except for homonymy.
Cucumis melo Linnaeus *parvus* C. Harz, Landw. Samenk. 2: 778. 1885. Published without indication of rank, so inoperative for priority except for homonymy.
Cucumis melo Linnaeus *pedatifidus* Alefeld, Landw. Fl. 199. 1866. Published without indication of rank, so inoperative for priority except for homonymy.
Cucumis melo Linnaeus var. *persicodorus* Seiz., Verh. Ver. Bef. Gartenb. 379. 1827.
Cucumis melo Linnaeus *persicus* C. Harz, Landw. Samenk. 2: 783. 1885. Published without indication of rank, so inoperative for priority except for homonymy.
Cucumis melo Linnaeus *peruanus* Alefeld, Landw. Fl. 203. 1866. Published without indication of rank, so inoperative for priority except for homonymy.
Cucumis melo Linnaeus *petrosus* Alefeld, Landw. Fl. 205. 1866. Published without indication of rank, so inoperative for priority except for homonymy.
Cucumis melo Linnaeus *polignae* Alefeld, Landw. Fl. 205. 1866. Published without indication of rank, so inoperative for priority except for homonymy.
Cucumis melo Linnaeus *portugalensis* C. Harz, Landw. Samenk. 2: 778. 1885. Published without indication of rank, so inoperative for priority except for homonymy.
Cucumis melo Linnaeus var. *praecantalupa* Pangalo in Zhukovsky, La Turquie agricole 537. 1933.
Cucumis melo Linnaeus *praecox* Alefeld, Landw. Fl. 204. 1866. Published without indication of rank, so inoperative for priority except for homonymy.
Cucumis melo Linnaeus *praecox* C. Harz, Landw. Samenk. 2: 779. 1885. Illegitemate, later homonym.
Cucumis melo Linnaeus var. *praecox* Filov, Vestn. Skh. Nauki 1: 130. 1960. Illegitimate, later homonym of *C. melo praecox* Alefeld since there was no direct reference to it by Filov.
Cucumis melo Linnaeus *prescotti* Alefeld, Landw. Fl. 207. 1866. Published without indication of rank, so inoperative for priority except for homonymy.
Cucumis melo Linnaeus *prescotti* C. Harz, Landw. Samenk. 2: 778. 1885. Published without indication of rank, so inoperative for priority except for homonymy.
Cucumis melo Linnaeus *provincialis* Alefeld, Landw. Fl. 202. 1866. Published without indication of rank, so inoperative for priority except for homonymy.
Cucumis melo Linnaeus subsp. *pubescens* (Willdenow) Hassib, Cucurbit. Egypt 86. 1938.—BASIONYM: *Cucumis pubescens* Willdenow.
Cucumis melo Linnaeus var. *pubescens* (Willdenow) Kurz, Journ. Asiat Soc. Beng. 46(2): 103. 1877.—BASIONYM: *Cucumis pubescens* Willdenow.
Cucumis melo Linnaeus *reginae* Alefeld, Landw. Fl. 199. 1866. Published without indication of rank, so inoperative for priority except for homonymy.
Cucumis melo Linnaeus var. *reticulatus* Chakravarty, Iraq Minist. Agric. Tech. Bull. 3: 95. 1977.—TYPE: IRAQ. Samarra, 24 Sep 1961, *Chakravarty & Nuri 30737* (holotype: BAG). Illegitimate, later homonym.
Cucumis melo Linnaeus var. *reticulatus* Naudin, Ann. Sc. Nat. Ser. 4, 11: 50. 1859. Illegitimate, later homonym.
Cucumis melo Linnaeus var. *reticulatus* Seringe in de Candolle, *Prodr*. 3: 300. 1828.
Cucumis melo Linnaeus var.-gr. *reticulatus* (Seringe) Alefeld, Landw. Fl. 363. 1866.—BASIONYM: *Cucumis melo* Linnaeus var. *reticulatus* Seringe in de Candolle.
Cucumis melo Linnaeus *rhodosarx* Alefeld, Landw. Fl. 205. 1866. Published without indication of rank, so inoperative for priority except for homonymy.
Cucumis melo Linnaeus *rigidus* Pangalo, Deutsche Landw. Rundschau 2: 62. 1928. Published without indication of rank, so inoperative for priority except for homonymy.
Cucumis melo Linnaeus subsp. *rigidus* (Pangalo) Filov, Vestn. Skh. Nauki 1: 130. 1960.—BASIONYM: *Cucumis melo* Linnaeus *rigidus* Pangalo.
Cucumis melo Linnaeus var. *rigidus* Pangalo, Trudy Prikl. Bot. 23(3): 245. 1930.
Cucumis melo Linnaeus *robusti* Pangalo, Russ. Bot. Congress Leningrad, Jan 1928, 301. 1928. Published without indication of rank, so inoperative for priority except for homonymy.
Cucumis melo Linnaeus *rogersii* C. Harz, Landw. Samenk. 2: 781. 1885. Published without indication of rank, so inoperative for priority except for homonymy.
Cucumis melo Linnaeus var. *rokkiford* Filov, Vestn. Skh. Nauki 1: 131. 1960. Invalid, type not designated.
Cucumis melo Linnaeus *romanus* Alefeld, Landw. Fl. 208. 1866. Published without indication of rank, so inoperative for priority except for homonymy.
Cucumis melo Linnaeus f. *rotundatus* Chakravarty, Iraq Minist. Agric. Tech. Bull. 3: 92. 1977.—TYPE: IRAQ. Kochek, 8 Oct 1961, *Chakravarty & Nuri 30741* (holotype: BAG).

Cucumis melo Linnaeus var. *rotundus* Chakravarty, Iraq Minist. Agric. Tech. Bull. 3: 92. 1977.—TYPE: IRAQ. Bejoari, 15 km south of Mosul, 9 Oct 1961, *Chakravarty & Nuri 30722* (holotype: BAG).

Cucumis melo Linnaeus f. *rugosus* Nakai ex Kitamura, Acta Phytotax. Geobot. 14: 43. 1950.

Cucumis melo Linnaeus var. *rugosus* Chakravarty, Iraq Minist. Agric. Tech. Bull. 3: 89. 1977.—TYPE: IRAQ. Bejoari, 8 km north of Mosul, 9 Oct 1961, *Chakravarty & Nuri 30723* (holotype: BAG). Illegitemate, later homonym

Cucumis melo Linnaeus *saccharatus* C. Harz, Landw. Samenk. 2: 782. 1885. Published without indication of rank, so inoperative for priority except for homonymy.

Cucumis melo Linnaeus var. *saccharinus* H. Jacquin, Monogr. melon 157. 1832.

Cucumis melo Linnaeus var. *saharunporensis* Naudin, Ann. Sc. Nat. Ser. 5, 5: 11. 1866.—TYPE (here designated): FRANCE. Cultivated at the Muséum d'Histoire Naturelle, seeds from Saharunpore, India, sent by Henry Cape, surgeon in the British army, 1863, *Naudin s.n.* (lectotype: P; isolectotype: P).

Cucumis melo Linnaeus var. *saidi* Hassib, Cucurbit. Egypt 124. 1938. Invalid, lacks description or diagnosis in Latin.

Cucumis melo Linnaeus *salmoneus* Alefeld, Landw. Fl. 206. 1866. Published without indication of rank, so inoperative for priority except for homonymy.

Cucumis melo Linnaeus var. *samarrensis* Chakravarty, Iraq Minist. Agric. Tech. Bull. 3: 93. 1977.—TYPE: IRAQ. Samarra, 29 Aug 1961, *Chakravarty & Nuri 30736* (holotype: BAG).

Cucumis melo Linnaeus var. *santawi* Hassib, Cucurbit. Egypt 126. 1938. Invalid, lacks description or diagnosis in Latin.

Cucumis melo Linnaeus *sapidissimus* Alefeld, Landw. Fl. 205. 1866. Published without indication of rank, so inoperative for priority except for homonymy.

Cucumis melo Linnaeus *schraderianus* Alefeld, Landw. Fl. 199. 1866. Published without indication of rank, so inoperative for priority except for homonymy.

Cucumis melo Linnaeus var. *senani* Hassib, Cucurbit. Egypt 125. 1938. Invalid, lacks description or diagnosis in Latin.

Cucumis melo Linnaeus var. *shahd* Hassib, Cucurbit. Egypt 128. 1938. Invalid, lacks description or diagnosis in Latin.

Cucumis melo Linnaeus var. *shammam* Saidi, Cucurbit. Egypt 130. 1938. Invalid, lacks description or diagnosis in Latin.

Cucumis melo Linnaeus var. *shauki* Chakravarty, Iraq Minist. Agric. Tech. Bull. 3: 91. 1977.—TYPE: IRAQ. Bejoari, 15 km south of Mosul, 9 Oct 1961, *Chakravarty & Nuri 30721* (holotype: BAG).

Cucumis melo Linnaeus var. *shimmam* Chakravarty, Iraq Minist. Agric. Tech. Bull. 3: 96. 1977.—TYPE: IRAQ. Sangar Chebier, 29 Oct 1961, *Chakravarty & Nuri 30728* (holotype: BAG).

Cucumis melo Linnaeus f. *showamelon* Kitamura, Acta Phytotax. Geobot. 14: 43. 1950.

Cucumis melo Linnaeus *siamensis* Alefeld, Landw. Fl. 207. 1866. Published without indication of rank, so inoperative for priority except for homonymy.

Cucumis melo Linnaeus *smyrnaeus* Sickenberger, Mem. Inst. Eqypt. 4(2): 234. 1901. Published without indication of rank, so inoperative for priority except for homonymy.

Cucumis melo Linnaeus subsp. *spontaneum* Filov, Vestn. Skh. Nauki 1: 131. 1960. Invalid, type not designated.

Cucumis melo Linnaeus var. *striata* (Sickenberger) Hassib, Cucurbit. Egypt 114. 1938.—BASIONYM: *Cucumis dudaim* Linnaeus *striata* Sickenberger.

Cucumis melo Linnaeus var. *suavis* Chakravarty, Iraq Minist. Agric. Tech. Bull. 3: 95. 1977.—TYPE: IRAQ. Baghdad river side, *Chakravarty & Nuri 30735* (holotype: BAG).

Cucumis melo Linnaeus *succosus* Alefeld, Landw. Fl. 208. 1866. Published without indication of rank, so inoperative for priority except for homonymy.

Cucumis melo Linnaeus *susianuus* Alefeld, Landw. Fl. 206. 1866. Published without indication of rank, so inoperative for priority except for homonymy.

Cucumis melo Linnaeus *syriacus* Alefeld, Landw. Fl. 205. 1866. Published without indication of rank, so inoperative for priority except for homonymy.

Cucumis melo Linnaeus var. *tamago* Makino in Iinuma, Somoku-Dzusetsu 4(2): 1309. 1912.—TYPE (here designated): *T. 41*, Somoku-Dzusetsu, vol. 4(2). 1912.

Cucumis melo Linnaeus var. *tarra* (Pangalo) Filov, Vestn. Skh. Nauki 1: 131. 1960. Invalid, lacks full and direct reference to place of valid publication of its basionym.

Cucumis melo Linnaeus *tauricus* Alefeld, Landw. Fl. 206. 1866. Published without indication of rank, so inoperative for priority except for homonymy.

Cucumis melo Linnaeus var. *texanus* Naudin, Ann. Sc. Nat. Ser. 4, 16: 160. 1862.—TYPE (here designated): FRANCE. Cultivated at the Muséum d'Histoire Naturelle, seeds from Philadelphia by Elias Durand, 1861, *Naudin s.n.* (lectotype: P; isolectotype: P).

Cucumis melo Linnaeus *theioides* Alefeld, Landw. Fl. 200. 1866. Published without indication of rank, so inoperative for priority except for homonymy.

Cucumis melo Linnaeus *thessalonicus* Alefeld, Landw. Fl. 203. 1866. Published without indication of rank, so inoperative for priority except for homonymy.

Cucumis melo Linnaeus *tubaeformis* C. Harz, Landw. Samenk. 2: 786. 1885. Published without indication of rank, so inoperative for priority except for homonymy.

114 Appendixes

Cucumis melo Linnaeus *turcicus* C. Harz, Landw. Samenk. 2: 778. 1885. Published without indication of rank, so inoperative for priority except for homonymy.
Cucumis melo Linnaeus *turonensis* Alefeld, Landw. Fl. 208. 1866. Published without indication of rank, so inoperative for priority except for homonymy.
Cucumis melo Linnaeus var. *tuzensis* Chakravarty, Iraq Minist. Agric. Tech. Bull. 3: 93. 1977.—TYPE: IRAQ. Suati Aksu, 11 Oct 1961, *Chakravarty & Nuri 30727* (holotype: BAG).
Cucumis melo Linnaeus f. *typicus* Pangalo, Trudy Prikl. Bot. 23(3): 245. 1930. It is the typical variety of *Cucumis melo* Linnaeus subsp. *gracilior* Pangalo.
Cucumis melo Linnaeus *unionis* Alefeld, Landw. Fl. 202. 1866. Published without indication of rank, so inoperative for priority except for homonymy.
Cucumis melo Linnaeus *unionis* C. Harz, Landw. Samenk. 2: 784. 1885. Published without indication of rank, so inoperative for priority except for homonymy.
Cucumis melo Linnaeus var. *utilissimus* (Roxburgh) Duthie & Fuller, Field Gard. Crops 2: 55. 1883.—BASIONYM: *Cucumis utilissimus* Roxburgh.
Cucumis melo Linnaeus ser. var. *vard* Gabaiev in Zhukovsky, La Turquie Agicole 540 1933.
Cucumis melo Linnaeus f. *variegatus* Makino, Bot. Mag. (Tokyo) 20: 81. 1906.
Cucumis melo Linnaeus var. *variegatus* Pangalo in Zhukovsky, La Turquie agricole 537. 1933.
Cucumis melo Linnaeus var. *varigatus* Chakravarty, Iraq Minist. Agric. Tech. Bull. 3: 92. 1977.—TYPE: IRAQ. *Chakravarty s.n.* (holotype: BAG). The main entry for this epithet is "varilgatus," and in the key it is "varigatus." Obviously the main entry is a typographical error.
Cucumis melo Linnaeus var. *vaughans-original-osage* Hassib, Cucurbit. Egypt 127. 1938. Invalid, lacks description or diagnosis in Latin.
Cucumis melo Linnaeus var. *virgatus* Chakravarty, Iraq Minist. Agric. Tech. Bull. 3: 95. 1977.—TYPE: IRAQ. Samarra, 24 Sep 1961, *Chakravarty & Nuri 30737* (holotype: BAG).
Cucumis melo Linnaeus *viridis* Alefeld, Landw. Fl. 207. 1866. Published without indication of rank, so inoperative for priority except for homonymy.
Cucumis melo Linnaeus *viridis* C. Harz, Landw. Samenk. 2: 780. 1885. Illegitemate, later homonym.
Cucumis melo Linnaeus f. *viridis* Makino, Bot. Mag. (Tokyo) 20: 81. 1906. Illegitimate, later homonym of *C. melo viridis* Alefeld since there was no direct reference to it by Makino.
Cucumis melo Linnaeus var. *viridis* Pangalo in Zhukovsky, La Turquie agricole 539. 1933. Illegitimate, later homonym of *C. melo viridis* Alefeld since there was no direct reference to it by Pangalo.
Cucumis melo Linnaeus subsp. *vulgaris* (H. Jacquin) Pangalo, Trudy Prikl. Bot. 23(3): 231, 250. 1930.—BASIONYM: *Cucumis melo* Linnaeus var. *vulgaris* H. Jacquin.
Cucumis melo Linnaeus var. *vulgaris* H. Jacquin, Monogr. melon, p. 143. 1832.
Cucumis melo Linnaeus subvar. *waraqi* Hassib, Cucurbit. Egypt 112. 1938. Invalid, lacks description or diagnosis in Latin.
Cucumis melo Linnaeus var. *zaami* Suzuki, Foundation of Hothouse Melon Culture 37. 1970. Nom. nud.
Cucumis melo Linnaeus convar. *zard* (Pangalo) Grebenscikov, Kulturpflanze 1: 136. 1953. Invalid, lacks full and direct reference to place of valid publication of its basionym.
Cucumis melo Linnaeus proles *zard* Pangalo, Melon cultivation in U.S.S.R. 185. 1934.
Cucumis melo Linnaeus ser. var. *zard* Gabaiev in Zhukovsky, La Turquie Agricole 540. 1933.
Cucumis melo Linnaeus var. *zebrino-aurantiacus* Pangalo in Zhukovsky, La Turquie agricole 540. 1933.
Cucumis melo Linnaeus var. *zebrino-luteus* Pangalo in Zhukovsky, La Turquie agricole 539. 1933.
Cucumis melo Linnaeus var. *zhukovskyi* Pangalo in Zhukovsky, La Turquie agricole 540. 1933.
Cucumis melo Linnaeus var. *zhukowskii* (Pangalo) Filov, Vestn. Skh. Nauki 1: 131. 1960.—BASIONYM: *Melo cassaba* Pangalo nid. *zhukowskii* Pangalo. Invalid, its basionym is invalid.
Cucumis microcarpus (Alefeld) Pangalo in Zhukovsky, La Turquie Agricole 534 1933.—BASIONYM: *Cucumis melo* Linnaeus *microcarpus* Alefeld.
Cucumis microsperma Nakai, Rigakukai 13(9): 14. 1916.
Cucumis microspermus Nakai var. *koreana* Nakai in T. Mori, Enum. pl. Corea 335. 1922. Nom. nud.
Cucumis microsperma Nakai var. *rugosa* Nakai, Rigakukai 13(9): 14. 1916.
Cucumis momordica Roxburgh, Fl. ind. *ed. 1832* 3: 720. 1832.—TYPE (here designated): Icones Roxburghianae *t. 456* (neotype: K!).
Cucumis moschatus S.F. Gray, Nat. arr. Brit. pl. 2: 552. 1821. Illegitimate, *Cucumis melo* Linnaeus cited as synonym.
Cucumis odoratissimus Moench, Methodus 654. 1794. Illegitimate, *C. dudaim* cited as synonym.
Cucumis officinarum-melo Crantz, Inst. rei herb. 1: 172. 1766.
Cucumis pancheranus Naudin, Ann. Sc. Nat. Bot., sér. 4, 12: 112, *t. 8*. 1859.—TYPE (here designated): FRANCE. Muséum d'Histoire Naturelle, grown from seed sent by J.A.I. Pancher from New Caledonia, 1859, *C.V. Naudin s.n.* (lectotype: K).
Cucumis pedatifidus Schrader, Linnaea 12: 418. 1838.
Cucumis persicus (Sageret) M. Roemer, Fam. nat, syn. monogr. 2: 71. 1846.—BASIONYM: *Melo persicus* Sageret.
Cucumis picrocarpus F. von Mueller, Trans. Phil. Inst. Vict. 3: 46. 1859.—TYPE: AUSTRALIA. In many parts of tropical Australia, *F. von Mueller* (Telford, 1982).

Cucumis pictus N. Jacquin, Hort. bot. vindob. 3: 17, *t. 27.* 1776.—TYPE (here designated: *T. 27, Hort. bot. vindob.,* vol. 3 (lectotype).

Cucumis princeps Wenderoth, Linnaea 15(Litt.): 100. 1841. *Cucumis pseudocolocynthis* Royle, Ill. bot. Himal. Mts.1: 220, 2: *t. 47, 2* and *b.* 1839.—TYPE (here designated): INDIA. In the open plains near Saharunpore and Delhi, *J.F. Royle Ill. bot. Himal. Mts.* 2: *t. 47, 2* (lectotype). Harrison (1978) cited the type specimen of *Cucumis pseudocolocynthis* Royle as being in Royle's herbarium now deposited at LIV. LIV sent on loan *Royle S474* annotated as *C. pseudocolocynthis.* Presumbly this is the specimen cited by Harrison. It has trifoliate leaves with stipules and stipels. It is definitely not a member of the genus *Cucumis,* and most probably is a member of the Fabaceae.

Cucumis pubescens Willdenow, Sp. pl. 4: 614. 1805.—TYPE: GERMANY. Plant cultivated at Berlin, Germany, *C.L. Willdenow s.n.* (holotype: B-W [IDC microfiche 7440, specimen number 18048!]).

Cucumis pyriformis Roxburgh ex Wight & Arnott, Prodr. fl. Ind. orient. 1: 342. 1834. Invalid, published as a synonym of *Cucumis turbinatus* Roxburgh.

Cucumis reflexus Zeih ex Seringe in de Candolle, Prodr. 3: 300. 1828. Invalid, published as synonym of *C. flexuosus* Linnaeus var. *reflexus* Seringe in de Candolle.

Cucumis reginae Schrader, Linnaea 12: 419. 1838.

Cucumis reticulatus Hortulanorum ex Steudel, Nomencl. *ed. 2,* 1(4): 451. 1840. Invalid, published as synonym of *C. melo* Linnaeus.

Cucumis saccharinus Hortulanorum ex Steudel, Nomencl. *ed. 2,* 1(4): 451. 1840. Invalid, published as synonym of *C. melo* Linnaues.

Cucumis schraderianus M.J. Roemer, Prospect syn. monogr. 2: 73. 1846.

Cucumis serotinus Haberle ex Seiz., Verh. Ver. Bef. Gartenb. 376. 1827.

Cucumis trigonus Roxburgh, Fl. ind. *ed. 1832* 1: 722. 1832.—TYPE (here designated): INDIA. *W. Roxburgh s.n.* (lectotype: K!).

Cucumis turbinatus Roxburgh, Fl. ind. *ed. 1832* 3: 723. 1832.—TYPE (here designated): Icones Roxburghianae *t. 464* (neotype: K!).

Cucumis umbilicatus R.A. Salisbury, Prodr. stirp. Chap. Allerton 157. 1796. Illegitimate, *C. melo* Linnaues cited in synonym.

Cucumis utilissimus Roxburgh, Fl. ind. *ed. 1832* 3: 721. 1832.—TYPE (here designated): Icones Roxburghianae *t. 462* (neotype: K!).

Cucumis verrucosus Hortulanorum ex Steudel, Nomencl. bot. *ed. 1,* 1: 244. 1821. Invalid, published as synonym of *C. melo* Linnaeus.

Cucumis villosus Boissier & Noé in Boissier, Diagn. pl. orient., sér. 2, 2: 59. 1856. Invalid, published as synonym of *C. eriocarpus* Boissier & Noé in Boissier.

Cucumis viridis Hortulanorum ex Steudel, Nomencl. bot. *ed. 2,* 1(4): 451. 1840. Invalid, published as synonym of *C. melo* Linnaeus.

Cucurbita aspera Solander ex Forster, Fl. ins. austr. 92. 1786. Nom. nud. (Cogniaux & Harms, 1924).

Melo adana Pangalo, Bot. Zhurn. (Moscow & Leningrad) 35: 576. 1950.

Melo adana nid. *complacentatus* Pangalo, Bot. Zhurn. (Moscow & Leningrad) 35(6): 576. 1950. Invalid, lacks Latin description or diagnosis.

Melo adana nid. *liquidoplacentatus* Pangalo, Bot. Zhurn. (Moscow & Leningrad) 35(6): 576. 1950. Invalid, lacks Latin description or diagnosis.

Melo adana nid. *trivialis* Pangalo, Bot. Zhurn. (Moscow & Leningrad) 35(6): 576. 1950. Invalid, lacks Latin description or diagnosis.

Melo adzhur Pangalo, Bot. Zhurn. (Moscow & Leningrad) 35: 575. 1950.

Melo adzhur nid. *curvatus* Pangalo, Bot. Zhurn. (Moscow & Leningrad) 35(6): 576. 1950. Invalid, lacks Latin description or diagnosis.

Melo adzhur nid. *subclavatus* Pangalo, Bot. Zhurn. (Moscow & Leningrad) 35(6): 576. 1950. Invalid, lacks Latin description or diagnosis.

Melo agrestis (Naudin) Pangalo, Bot. Zhurn. (Moscow & Leningrad) 35: 580. 1950.—BASIONYM: *Cucumis melo* Linnaeus var. *agrestis* Naudin.

Melo X ambiguua Pangalo, Bot. Zhurn. (Moscow & Leningrad) 35: 577. 1950. Parents are given as *M. cantalupa* X *cassaba.*

Melo ameri Pangalo, Bot. Zhurn. (Moscow & Leningrad) 35: 578. 1950.

Melo ameri Pangalo nid. *albidus* Pangalo, Bot. Zhurn. (Moscow & Leningrad) 35(6): 578. 1950. Invalid, lacks Latin description or diagnosis.

Melo ameri Pangalo nid. *ananas* Pangalo, Bot. Zhurn. (Moscow & Leningrad) 35(6): 579. 1950. Invalid, lacks Latin description or diagnosis.

Melo ameri Pangalo nid. *bargi* Pangalo, Bot. Zhurn. (Moscow & Leningrad) 35(6): 579. 1950. Invalid, lacks Latin description or diagnosis.

Melo ameri Pangalo nid. *chiar* Pangalo, Bot. Zhurn. (Moscow & Leningrad) 35(6): 579. 1950. Invalid, lacks Latin description or diagnosis.

Melo ameri Pangalo nid. *chtai* Pangalo, Bot. Zhurn. (Moscow & Leningrad) 35(6): 579. 1950. Invalid, lacks Latin description or diagnosis.

Melo ameri Pangalo nid. *maculatus* Pangalo, Bot. Zhurn. (Moscow & Leningrad) 35(6): 579. 1950. Invalid, lacks Latin description or diagnosis.
Melo ameri Pangalo nid. *wacharman* Pangalo, Bot. Zhurn. (Moscow & Leningrad) 35(6): 579. 1950. Invalid, lacks Latin description or diagnosis.
Melo cantalupensis (Naudin) Pangalo, Bot. Zhurn. (Moscow & Leningrad) 35: 576. 1950. BASIONYM: *Cucumis melo* Linnaeus var. *cantalupensis* Naudin.
Melo cantalupensis (Naudin) Pangalo nid. *algeriensis* Pangalo, Bot. Zhurn. (Moscow & Leningrad) 35(6): 577. 1950. Invalid, lacks Latin description or diagnosis.
Melo cantalupensis (Naudin) Pangalo nid. *banana* Pangalo, Bot. Zhurn. (Moscow & Leningrad) 35(6): 577. 1950. Invalid, lacks Latin description or diagnosis.
Melo cantalupensis (Naudin) Pangalo nid. *carmelitanus* Pangalo, Bot. Zhurn. (Moscow & Leningrad) 35(6): 577. 1950. Invalid, lacks Latin description or diagnosis.
Melo cantalupensis (Naudin) Pangalo nid. *compactus* Pangalo, Bot. Zhurn. (Moscow & Leningrad) 35(6): 577. 1950. Invalid, lacks Latin description or diagnosis.
Melo cantalupensis (Naudin) Pangalo nid. *compressus* Pangalo, Bot. Zhurn. (Moscow & Leningrad) 35(6): 577. 1950. Invalid, lacks Latin description or diagnosis.
Melo cantalupensis (Naudin) Pangalo nid. *primigenius* Pangalo, Bot. Zhurn. (Moscow & Leningrad) 35(6): 576. 1950. Invalid, lacks Latin description or diagnosis.
Melo cantalupensis (Naudin) Pangalo nid. *turensis* Pangalo, Bot. Zhurn. (Moscow & Leningrad) 35(6): 577. 1950. Invalid, lacks Latin description or diagnosis.
Melo cassaba Pangalo, Bot. Zhurn. (Moscow & Leningrad) 35: 575. 1950.
Melo cassaba Pangalo nid. *adress* Pangalo, Bot. Zhurn. (Moscow & Leningrad) 35(6): 576. 1950. Invalid, lacks Latin description or diagnosis.
Melo cassaba Pangalo nid. *aureus* Pangalo, Bot. Zhurn. (Moscow & Leningrad) 35(6): 576. 1950. Invalid, lacks Latin description or diagnosis.
Melo cassaba Pangalo nid. *burikala* Pangalo, Bot. Zhurn. (Moscow & Leningrad) 35(6): 576. 1950. Invalid, lacks Latin description or diagnosis.
Melo cassaba Pangalo nid. *candicans* Pangalo, Bot. Zhurn. (Moscow & Leningrad) 35(6): 576. 1950. Invalid, lacks Latin description or diagnosis.
Melo cassaba Pangalo nid. *gurgak* Pangalo, Bot. Zhurn. (Moscow & Leningrad) 35(6): 575. 1950. Invalid, lacks Latin description or diagnosis.
Melo cassaba Pangalo nid. *orientalis* Pangalo, Bot. Zhurn. (Moscow & Leningrad) 35(6): 576. 1950. Invalid, lacks Latin description or diagnosis.
Melo cassaba Pangalo nid. *zagara* Pangalo, Bot. Zhurn. (Moscow & Leningrad) 35(6): 575. 1950. Invalid, lacks Latin description or diagnosis.
Melo cassaba Pangalo nid. *zebrinus* Pangalo, Bot. Zhurn. (Moscow & Leningrad) 35(6): 576. 1950. Invalid, lacks Latin description or diagnosis.
Melo cassaba Pangalo nid. *zhukowskii* Pangalo, Bot. Zhurn. (Moscow & Leningrad) 35(6): 575. 1950. Invalid, lacks Latin description or diagnosis.
Melo chandalak Pangalo, Bot. Zhurn. (Moscow & Leningrad) 35: 577. 1950.
Melo chandalak Pangalo nid. *bucharicus* Pangalo, Bot. Zhurn. (Moscow & Leningrad) 35(6): 578. 1950. Invalid, lacks Latin description or diagnosis.
Melo chandalak Pangalo nid. *caram* Pangalo, Bot. Zhurn. (Moscow & Leningrad) 35(6): 578. 1950. Invalid, lacks Latin description or diagnosis.
Melo chandalak Pangalo nid. *garma* Pangalo, Bot. Zhurn. (Moscow & Leningrad) 35(6): 578. 1950. Invalid, lacks Latin description or diagnosis.
Melo chandalak Pangalo nid. *glaucus* Pangalo, Bot. Zhurn. (Moscow & Leningrad) 35(6): 578. 1950. Invalid, lacks Latin description or diagnosis.
Melo chandalak Pangalo nid. *imdamas* Pangalo, Bot. Zhurn. (Moscow & Leningrad) 35(6): 578. 1950. Invalid, lacks Latin description or diagnosis.
Melo chandalak Pangalo nid. *loreus* Pangalo, Bot. Zhurn. (Moscow & Leningrad) 35(6): 578. 1950. Invalid, lacks Latin description or diagnosis.
Melo chandalak Pangalo nid. *tarnak* Pangalo, Bot. Zhurn. (Moscow & Leningrad) 35(6): 578. 1950. Invalid, lacks Latin description or diagnosis.
Melo chandalak Pangalo nid. *tochmi* Pangalo, Bot. Zhurn. (Moscow & Leningrad) 35(6): 578. 1950. Invalid, lacks Latin description or diagnosis.
Melo chandalak Pangalo nid. *variegatus* Pangalo, Bot. Zhurn. (Moscow & Leningrad) 35(6): 578. 1950. Invalid, lacks Latin description or diagnosis.
Melo chandalak Pangalo nid. *zami* Pangalo, Bot. Zhurn. (Moscow & Leningrad) 35(6): 578. 1950. Invalid, lacks Latin description or diagnosis.
Melo chandalak Pangalo nid. *zamira* Pangalo, Bot. Zhurn. (Moscow & Leningrad) 35(6): 578. 1950. Invalid, lacks Latin description or diagnosis.
Melo chate Sageret ex M. Roemer, *Fam. nat. syn. monogr.* 2: 74. 1846. Invalid, published as synonym of *C. chate* Linnaeus.
Melo chate Sageret, Ann. Sci. Nat. 8: 312. 1826. Nom. nud.
Melo chate Sageret, Mem. Soc. Roy. et Centr. Agric. 58: 488. 1825. Nom. nud.

Melo chinensis Pangalo, Bot. Zhurn. (Moscow & Leningrad) 35: 580. 1950.

Melo conomon Pangalo, Bot. Zhurn. (Moscow & Leningrad) 35: 580. 1950.

Melo dudaim Sageret, Ann. Sc. Nat. 8: 312. 1826. Nom. nud.

Melo dudaim Sageret, Mem. Soc. Roy. et Centr. Agric. 58: 488. 1825. Nom. nud.

Melo figari Pangalo, Bot. Zhurn. (Moscow & Leningrad) 35(6): 580. 1950. Invalid,

Melo flexuosus Sageret, Ann. Sci. Nat. 8: 312. 1826. Nom. nud.

Melo flexuosus (Linnaeus) Pangalo, Bot. Zhurn. (Moscow & Leningrad) 35: 577. 1950.—BASIONYM: *Cucumis flexuosus* Linnaeus.

Melo flexuosus Sageret, Mem. Soc. Roy. et Centr. Aric. 58: 488. 1825. Nom. nud.

Melo flexuosus Sageret ex M. Roemer, Fam. nat. syn. Monogr. 2: 75. 1846. Invalid, published as synonym of *C. flexuosus* Linnaeus.

Melo microcarpus (Alefeld) Pangalo, Bot. Zhurn. (Moscow & Leningrad) 35(6): 574. 1950. BASIONYM: *Cucumis melo* Linnaeus *microcarpus* Alefeld.

Melo microcarpus (Alefeld) Pangalo nid. *albus* Pangalo, Bot. Zhurn. (Moscow & Leningrad) 35(6): 575. 1950. Invalid, lacks Latin description or diagnosis.

Melo microcarpus (Alefeld) Pangalo nid. *dudaim* Pangalo, Bot. Zhurn. (Moscow & Leningrad) 35(6): 575. 1950. Invalid, lacks Latin description or diagnosis.

Melo microcarpus (Alefeld) Pangalo nid. *erythraceus* Pangalo, Bot. Zhurn. (Moscow & Leningrad) 35(6): 575. 1950. Invalid, lacks Latin description or diagnosis.

Melo microcarpus (Alefeld) Pangalo nid. *flavus* Pangalo, Bot. Zhurn. (Moscow & Leningrad) 35(6): 575. 1950. Invalid, lacks Latin description or diagnosis.

Melo monoclinus Pangalo, Bot. Zhurn. (Moscow & Leningrad) 35: 580. 1950.

Melo orientalis (S. Kudr.) Nabiev, Fl. Uzbekist. v. 609. 1961. BASIONYM: *Cucumis orientalis* S. Kudr., Veg. Guzar. 142. 1941; original publication of basionym not examined.

Melo persicus Sageret, Mem. Soc. Roy. et Centr. Agric. 58: 488. 1825.

Melo sativus Sageret, Mem. Soc. Roy. et Centr. Agric. 58: 437, 488. 1825. New name for *Cucumis melo* Linnaeus when transferred to *Melo*.

Melo vulgaris Moench ex Cogniaux in de Candolle, Monogr. phan. 3: 484. 1881.

Melo zard Pangalo, Bot. Zhurn. (Moscow & Leningrad) 35: 579. 1950.

Melo zard Pangalo nid. *asma* Pangalo, Bot. Zhurn. (Moscow & Leningrad) 35(6): 579. 1950. Invalid, lacks Latin description or diagnosis.

Melo zard Pangalo nid. *aurantiacus* Pangalo, Bot. Zhurn. (Moscow & Leningrad) 35(6): 579. 1950. Invalid, lacks Latin description or diagnosis.

Melo zard Pangalo nid. *gulabus* Pangalo, Bot. Zhurn. (Moscow & Leningrad) 35(6): 579. 1950. Invalid, lacks Latin description or diagnosis.

Melo zard Pangalo nid. *hibernus* Pangalo, Bot. Zhurn. (Moscow & Leningrad) 35(6): 580. 1950. Invalid, lacks Latin description or diagnosis.

Melo zard Pangalo nid. *kalassan* Pangalo, Bot. Zhurn. (Moscow & Leningrad) 35(6): 579. 1950. Invalid, lacks Latin description or diagnosis.

Melo zard Pangalo nid. *lapidosus* Pangalo, Bot. Zhurn. (Moscow & Leningrad) 35(6): 580. 1950. Invalid, lacks Latin description or diagnosis.

Melo zard Pangalo nid. *nokki* Pangalo, Bot. Zhurn. (Moscow & Leningrad) 35(6): 579. 1950. Invalid, lacks Latin description or diagnosis.

Melo zard Pangalo nid. *rugulosus* Pangalo, Bot. Zhurn. (Moscow & Leningrad) 35(6): 579. 1950. Invalid, lacks Latin description or diagnosis.

Melo zard Pangalo nid. *solidus* Pangalo, Bot. Zhurn. (Moscow & Leningrad) 35(6): 579. 1950. Invalid, lacks Latin description or diagnosis.

Melo zard Pangalo nid. *tarlama* Pangalo, Bot. Zhurn. (Moscow & Leningrad) 35(6): 580. 1950. Invalid, lacks Latin description or diagnosis.

Melo zard Pangalo nid. *tenebrosus* Pangalo, Bot. Zhurn. (Moscow & Leningrad) 35(6): 580. 1950. Invalid, lacks Latin description or diagnosis.

Appendix 2. Synonyms of *Cucumis sativus* Linnaeus

Cucumis esculentus R.A. Salisbury, Prodr. stirp. Chap. Allerton 157. 1796. Illegitimate, *C. sativus* Linnaues given in synonymy.
Cucumis hardwickii Royle, Ill. bot. Himal. Mts. 1: 220, *t. 47, 3* and *a*. 1835.—TYPE: INDIA. Northwestern India, J.F. Royle s.n. (holotype: LIV!; isotypes: K! LE!).
Cucumis muricatus Willdenow, Sp. pl. 4: 613. 1805.—TYPE: Tranquebar, D. Klein s.n..
Cucumis rumphii Hasskarl, Neu. Schlüss. Rumph. 280 (Merrill, 1917). 1866.
Cucumis sativus Linnaeus subsp. *agrestis* Gabaev, Cucumbers 35 [in Russian]. 1932.
Cucumis sativus Linnaeus f. *albus* Hiroe, Pl. Basho's & Buson's Hokku Lit. 8(3): 191, *f. 3*. 1973.
Cucumis sativus Linnaeus f. *albus* Pangalo in Zhukovsky, La Turquie agricole 557, 882. 1933.
Cucumis sativus Linnaeus var. *albus* Seringe in de Candolle, Prodr. 3: 300. 1828.
Cucumis sativus Linnaeus var. *anatolicus* Gabaev, Trudy Prikl. Bot. 23(3): 452, *f. 20 & 22*. 1930.
Cucumis sativus Linnaeus var. *anglicus* Bailey, Stand. cycl. hort. 2: 908, *f. 1127*. 1914.
Cucumis sativus Linnaeus *antasiaticus* Gabaev, Trudy Prikl. Bot. 23(3): 452. 1930. Invalid, author stated intent not to publish taxonomically.
Cucumis sativus Linnaeus var. *arakis* Forsskål, Fl. aegypt.-arab. 169. 1775.
Cucumis sativus Linnaeus f. *australis* Kitamura, Acta Phytotax. Geobot. 14: 44. 1950.
Cucumis sativus Linnaeus var. *battich-djebbal* Forsskål, Fl. aegypt.-arab. 169. 1775.
Cucumis sativus Linnaeus f. *borealis* Kitamura, Acta Phytotax. Geobot. 14: 44. 1950.
Cucumis sativus Linnaeus var. *brullos* Forsskål, Fl. aegypt.-arab. 169. 1775.
Cucumis sativus Linnaeus f. *brunnescens* Gabaev, Trudy Prikl. Bot. 23(3): 455. 1930. Simultaneous homonym of *C. sativus* Linnaeus f. *brunnescens* Gabaev, page 456.
Cucumis sativus Linnaeus f. *brunnescens* Gabaev, Trudy Prikl. Bot. 23(3): 456. 1930. Simultaneous homonym of *C. sativus* Linnaeus f. *brunnescens* Gabaev, page 455.
Cucumis sativus Linnaeus var. *chatte* Forsskål, Fl. aegypt.-arab. 169. 1775.
Cucumis sativus Linnaeus var. *chiar* Forsskål, Fl. aegypt.-arab. 169. 1775.
Cucumis sativus Linnaeus var. *cilicicus* Gabaev, Trudy Prikl. Bot. 23(3): 452, 465, *f. 22 & 24*. 1930.
Cucumis sativus Linnaeus *curtus* Alefeld, Landw. Fl. 196. 1866. Published without indication of rank, so inoperative for priority except for homonymy.
Cucumis sativus Linnaeus *donii* Alefeld, Landw. Fl. 197. 1866. Published without indication of rank, so inoperative for priority except for homonymy.
Cucumis sativus Linnaeus var. *ennemis* Forsskål, Fl. aegypt.-arab. 169. 1775.
Cucumis sativus Linnaeus var. *europaeus* Gabaev, Cucumbers 35, 39, 62. 1932.
Cucumis sativus Linnaeus *excellens* Alefeld, Landwirtsch. Flora 196. 1866. Published without indication of rank, so inoperative for priority except for homonymy.
Cucumis sativus Linnaeus var. *fakus* Forsskål, Fl. aegypt.-arab. 169. 1775.
Cucumis sativus Linnaeus var. *falcatus* Gabaev, Trudy Prikl. Bot. 23(3): 451, 454, *f. 10, 11, & 12*. 1930.
Cucumis sativus Linnaeus var. *fastigiatus* Seringe in de Candolle, Prodr. 3: 300. 1828.
Cucumis sativus Linnaeus var. *flavus* Seringe in de Candolle, Prodr. 3: 300. 1828.
Cucumis sativus Linnaeus *flexuosus* Alefeld, Landw. Fl. 197. 1866. Published without indication of rank, so inoperative for priority except for homonymy.
Cucumis sativus Linnaeus *gracilior* Gabaev, Trudy Prikl. Bot. 23(3): 452. 1930. Invalid, author stated intent not to publish taxonomically.
Cucumis sativus Linnaeus subsp. *gracilior* Gabaev, Cucumbers 36 [in Russian]. 1932. This is referred to once in a table, page 35, as "*gracilis*," a presumed orthographic variant.
Cucumis sativus Linnaeus var. *grossularioides* Tkachenko, Trudy Prikl. Bot., Ser. 2, Genet. Rast. 9: 328. 1935. Invalid, lacks Latin description or diagnosis.
Cucumis sativus Linnaeus *hardwickii* (Royle) Alefeld, Landw. Fl. 196. 1866. Published without indication of rank, so inoperative for priority except for homonymy.
Cucumis sativus Linnaeus var. *hardwickii* (Royle) Gabaev, Cucumbers 47 [in Russian]. 1932.—BASIONYM: *Cucumis hardwickii* Royle.
Cucumis sativus Linnaeus *hollandicus* Alefeld, Landwirtsch. Flora 197. 1866. Published without indication of rank, so inoperative for priority except for homonymy.
Cucumis sativus Linnaeus var. *indo-europeus* Gabaev, Cucumbers 35, 43, 57. 1932.
Cucumis sativus Linnaeus var. *irano-turanicus* Gabaev, Trudy Prikl. Bot. 23(3): 452, 461, *f. 18 & 19*. 1930.
Cucumis sativus Linnaeus var. *izmir* Gabaev, Trudy Prikl. Bot. 23(3): 452, 463, *f. 23*. 1930.
Cucumis sativus Linnaeus *longus* Harz, Landw. Samenk. 2: 775. 1885. Published without indication of rank, so inoperative for priority except for homonymy.

Cucumis sativus Linnaeus *opheocarpus* Harz, Landw. Samenk. 2: 775. 1885. Published without indication of rank, so inoperative for priority except for homonymy.

Cucumis sativus Linnaeus *orasiaticus* Gabaev, Trudy Prikl. Bot. 23(3): 452. 1930. Invalid, author stated intent not to publish taxonomically.

Cucumis sativus Linnaeus f. *pallescens* Gabaev, Trudy Prikl. Bot. 23(3): 456. 1930. Simultaneous homonym of *C. sativus* Linnaeus f. *pallescens* Gabaev, page 455.

Cucumis sativus Linnaeus f. *pallescens* Gabaev, Trudy Prikl. Bot. 23(3): 455. 1930. Simeltaneous homonym of *C. sativus* Linnaeus f. *pallescens* Gabaev, page 456.

Cucumis sativus Linnaeus *pallidus* Alefeld, Landw. Fl. 197. 1866. Published without indication of rank, so inoperative for priority except for homonymy.

Cucumis sativus Linnaeus var. *pallidus* Gabaev in Zhukovsky, La Turquie agricole 552. 1933.

Cucumis sativus Linnaeus *praecox* Alefeld, Landw. Fl. 197. 1866. Published without indication of rank, so inoperative for priority except for homonymy.

Cucumis sativus Linnaeus subsp. *rigidus* Gabaev, Cucumbers 35. 1932.

Cucumis sativus Linnaeus subsp. *rigidus* Gabaev, Trudy Prikl. Bot. 23(3): 451, 571, 472. 1930. Invalid, author stated intent not to publish taxonomically.

Cucumis sativus Linnaeus *rossicus* Alefeld, Landw. Fl. 197. 1866. Published without indication of rank, so inoperative for priority except for homonymy.

Cucumis sativus Linnaeus var. *schemmam* Forsskål, Fl. aegypt.-arab. 169. 1775.

Cucumis sativus Linnaeus *serotinus* Alefeld, Landwirtsch. Flora 197. 1866. Published without indication of rank, so inoperative for priority except for homonymy.

Cucumis sativus Linnaeus *setosus* Alefeld, Landw. Fl. 197. 1866. Published without indication of rank, so inoperative for priority except for homonymy.

Cucumis sativus Linnaeus var. *sikkimensis* Hooker filius, Bot. mag. 192: *t. 6206.* 1876.—TYPES: EUROPE. Cultivated in the economic house from seed sent by T. Clarke, Aug 1875, *Royal Botanic Gardens, Kew, s.n.* (holotype: K!; isotype: K!). INDIA. Sikkim: Cultivated, detached fruits and seeds, *J.D Hooker s.n.* (paratype: K!).

Cucumis sativus Linnaeus *sikkimiae* Harz, Landw. Samenk. 775. 1885. Published without indication of rank, so inoperative for priority except for homonymy.

Cucumis sativus Linnaeus *smilli* Forsskål, Fl. aegypt.-arab. 169. 1775.

Cucumis sativus Linnaeus var. *squamosus* Gabaev, Trudy Prikl. Bot. 23(3): 451, 459, *f. 16.* 1930.

Cucumis sativus Linnaeus var. *testudaceus* Gabaev, Trudy Prikl. Bot. 23(3): 451, 459, *f. 15.* 1930.

Cucumis sativus Linnaeus f. *tuberculatus* Hiroe, Pl. Basho's & Buson's Hokku Lit. 8(3): 190, *f. 1-5.* 1973.

Cucumis sativus Linnaeus var. *tuberculatus* Gabaev, Trudy Prikl. Bot. 23(3): 451, 456, *f. 13.* 1930.

Cucumis sativus Linnaeus *turcicus* Alefeld, Landw. Fl. 197. 1866. Published without indication of rank, so inoperative for priority except for homonymy.

Cucumis sativus Linnaeus f. *typicus* Gabaev in Zhukovsky, La Turquie agricole 556. 1933. This is the typical form of *C. sativus* Linnaeus var. *cilicicus* Gabaev.

Cucumis sativus Linnaeus var. *variegatus* Seringe ex de Candolle, Prodr. 3: 300. 1828.

Cucumis sativus Linnaeus f. *viridis* Gabaev in Zhukovsky, La Turquie agricole 556. 1933. Illegitemate, later homonym.

Cucumis sativus Linnaeus f. *viridis* Gabaev in Zhukovsky, La Turquie agricole 558, 882. 1933. Illegitemate, later homonym.

Cucumis sativus Linnaeus var. *viridis* Seringe ex de Candolle, Prodr. 3: 300. 1828.

Cucumis sativus Linnaeus var.-gr. *viridis* (Seringe ex de Candolle) Alefeld, Landw. Fl. 196. 1866.

Cucumis sativus Linnaeus *vulgaris* Alefeld, Landw. Fl. 196. 1866. Published without indication of rank, so inoperative for priority except for homonymy.

Cucumis sativus Linnaeus var. *vulgatus* Gabaev, Trudy Prikl. Bot. 23(3): 451, 456, *f. 10, 11, & 14.* 1930.

Cucumis sativus Linnaeus var. *xishuangbannanesis* Qi Chunzhang & Yuan Zhenzhen, Acta Hort. Sin. 10(4): 259. 1983. Nom. nud.

Cucumis setosus Cogniaux in de Candolle, Monogr. phan. 3: 491. 1881.—TYPE: INDIA. Belgaum, *D. Ritchie 321* (holotype: E!; isotype: K!).

Cucumis sphaerocarpus Gabaev, Cucumbers 28, 78, *f. 58* [in Russian]. 1932.

Cucumis vilmorinii Sprenger, Bull. Soc. Tosc. Ortic. 19: 115. 1894.

#1. Plants <habit>/
    1. vines/
    2. herbs/
    3. subshrubs/
    4. shrubs/
#2. Plants <life cycle>/
    1. annual/
    2. perennial/
#3. Plants <condition of taproot>/
    1. without a woody rootstock/
    2. with a woody rootstock/
#4. Plants <tuberous or not>/
    1. lacking tubers/
    2. tuberous/
#5. Plants <reproductive strategy>/
    1. monoecious <male and female flowers on the same plant>/
    2. dioecious <male and female flowers on different plants>/
    3. andromonoecious <male and perfect flowers on the same plant>/
#6. Tubers <depth under ground in cm>/
    cm under ground/
#7. Tubers <shape>/
    1. ellipsoid/
    2. obovoid/
#8. Tubers <height in cm>/
    cm long/
#9. Tubers <diameter in cm>/
    cm in diam/
#10. Tubers <pubescence type>/
    1. glabrous/
    2. FILLER/
#11. Stems <orientation>/
    1. procumbent/
    2. climbing/
    3. erect/
#12. Stems <shape in transverse section>/
    1. sulcate <alternating longitudinal grooves and ridges>/
    2. terete <smooth circular in outline, cylindric>/
#13. Stems <aculeate or not>/
    1. not aculeate/
    2. aculeate/
#14. Stems <pubescence type>/
    1. glabrous/
    2. glabrate/
    3. hispid/
    4. hispidulous/
    5. villous/
    6. pilose/
    7. hirsute/
    8. antrorse-strigose/
    9. retrorse-strigose/
    10. scabrous/
    11. setose/
    12. lanate/
    13. puberulent/
#15. Stems <individual hair types>/
    1. with nonbreakaway hairs/
    2. with breakaway hairs <nonglandular, multicellular, conical hairs
        consisting of an easily ruptured, multicellular foot with

numerous, small, thin-walled cells and a uniseriate body with
larger, thick-walled cells (Inamdar & Gangadhara, 1975; Inamdar
et al., 1990)>/
#16. Stem hairs <length in mm>/
     mm long/
#17. Nodes <geniculate or not>/
     1. not geniculate/
     2. geniculate/
#18. Internodes <length in cm>/
     cm long/
#19. Petioles <length in cm>/
     cm long/
#20. Petioles <shape in outline>/
     1. sulcate in cross section <alternating longitudinal grooves and
       ridges in cross section>/
     2. terete in cross section <smooth circular in outline, cylindric>/
#21. Petioles <aculeate or not>/
     1. not aculeate/
     2. aculeate/
#22. Petioles <pubescence type uniform or 2/3 types on each petiole>/
     1. pubescence a single type on each petiole/
     2. pubescence 2 different types uniformly intermixed on each
       petiole/
     3. pubescence 3 different types in distinct zones on each petiole/
#23. Petioles <pubescence type>/
     1. glabrous/
     2. hispid/
     3. hispidulous/
     4. villous/
     5. pilose/
     6. lanate/
     7. hirsute/
     8. antrorse-strigose/
     9. retrorse-strigose/
     10. setose/
     11. scabrous/
#24. Petioles <individual hair type>/
     1. with nonbreakaway hairs/
     2. with breakaway hairs <nonglandular, multicellular, conical hairs
       consisting of an easily ruptured, multicellular foot with
       numerous, small, thin-walled cells and a uniseriate body with
       larger, thick-walled cells (Inamdar & Gangadhara, 1975; Inamdar
       et al., 1990)>/
#25. Leaf blades <lobation>/
     1. entire/
     2. trilobate <with 3 shallow lobes>/
     3. pentalobate <with 5 shallow lobes>/
     4. heptalobate <with 7 shallow lobes>/
     5. 3-palmately lobed <deeply divided into 3 lobes>/
     6. 5-palmately lobed <deeply divided into 5 lobes>/
#26. Leaf blades <condition of margin>/
     1. with the margin entire/
     2. with the margin serrate/
#27. Leaf blades <overall shape>/
     1. elliptic in outline <length to width ratio: 1.35-2.5>/
     2. narrowly ovate in outline <length to width ratio: 2.5-9.0>/
     3. ovate in outline <length to width ratio: 1.35-2.5>/
     4. broadly ovate in outline <length to width ratio: 1.0-1.35>/
     5. very broadly ovate in outline <length to width ratio: 0.75-1.0>/
     6. shallowly ovate in outline <length to width ratio: 0.4-0.75>/
     7. triangular in outline <length to width ratio: 1.35-2.5>/
#28. Leaf blades <shape of base>/
     1. acute at the base/
     2. broadly acute at the base/
     3. obtuse at the base/
     4. subtruncate at the base/

     5. truncate at the base/
     6. subcordate at the base/
     7. cordate at the base <lobes more or less obtuse>/
     8. sagittate at the base <lobes acute and oriented down or in same
        direction as petiole or midrib>/
     9. hastate at the base <lobes acute and pointed outwards at right
        angles to petiole or midrib>/
#29. Leaf blades <basal sinus, presence/absence>/
     1. with a basal sinus/
     2. without a basal sinus/
#30. Leaf blades <basal sinus, depth in cm>/
     cm deep/
#31. Leaf blades <shape of apex>/
     1. narrowly acute at the apex/
     2. acute at the apex/
     3. broadly acute at the apex/
     4. acuminate at the apex/
     5. broadly acuminate at the apex/
     6. obtuse at the apex/
#32. Leaf blades <overall length in cm>/
     cm/
#33. Leaf blades <overall width in cm>/
     cm/
#34. Leaf blades <length to width ratio>/
     times longer than wide/
#35. Leaf blades <pubescence type on upper surface>/
     1. hispid on the upper surface/
     2. hispidulous on the upper surface/
     3. pilose on the upper surface/
     4. hirsute on the upper surface/
     5. antrorse-strigose on the upper surface/
     6. scabrous on the upper surface/
#36. Leaf blades <vein on lower surface - aculeate or not>/
     1. not aculeate on the veins below/
     2. aculeate on the veins below/
#37. Leaf blades <vein pubescence type on lower surface>/
     1. glabrous on the veins below/
     2. hispid on the veins below/
     3. hispidulous on the veins below/
     4. pilose on the veins below/
     5. hirsute on the veins below/
     6. antrorse-strigose on the veins below/
     7. scabrous on the veins below/
#38. Leaf blades <intervenium pubescence type on lower surface>/
     1. glabrous on the intervenium below/
     2. hispid on the intervenium below/
     3. hispidulous on the intervenium below/
     4. hirsute on the intervenium below/
     5. pilose on the intervenium below/
     6. scabrous on the intervenium below/
     7. antrorse-strigose on the intervenium below/
#39. Leaf blades <individual hair types>/
     1. with nonbreakaway hairs/
     2. with breakaway hairs <nonglandular, multicellular, conical hairs
        consisting of an easily ruptured, multicellular foot with
        numerous, small, thin-walled cells and a uniseriate body with
        larger, thick-walled cells (Inamdar & Gangadhara, 1975; Inamdar
        et al., 1990)>/
#42. Central leaf-blade lobe <lobation>/
     1. entire/
     2. pinnatifid/
#43. Central leaf-blade lobe <shape>/
     1. narrowly elliptic in outline <length to width ratio: 2.5-9.0>/
     2. elliptic in outline <length to width ratio: 1.35-2.5>/
     3. broadly elliptic in outline <length to width ratio: 1.0-1.35>/
     4. narrowly linear in outline <length to width ratio: greater than 9.0>/

5. narrowly oblong in outline <length to width ratio: 2.5-9.0>/
6. narrowly ovate in outline <length to width ratio: 2.5-9.0>/
7. ovate in outline <length to width ratio: 1.35-2.5>/
8. broadly ovate in outline <length to width ratio: 1.0-1.35>/
9. very broadly ovate in outline <length to width ratio: 0.75-1.0>/
10. shallowly ovate in outline <length to width ratio: 0.4-0.75>/
11. very shallowly ovate in outline <length to width ratio: 0.11-0.4>/
12. narrowly triangular in outline <length to width ratio: 2.5-9.0>/
13. broadly triangular in outline <length to width ratio: 1.0-1.35>/
14. very broadly triangular in outline <length to width ratio: 0.75-1.0>/
15. shallowly triangular in outline <length to width ratio: 0.4-0.75>/
#44. Central leaf-blade lobe <shape at apex>/
1. narrowly acute at the apex/
2. acute at the apex/
3. broadly acute at the apex/
4. acuminate at the apex/
5. broadly acuminate at the apex/
6. obtuse at the apex/
#45. Central leaf-blade lobe <length in cm>/
cm/
#46. Central leaf-blade lobe <width in cm>/
cm/
#47. Lateral leaf-blade lobes <symmetry>/
1. symmetrical/
2. asymmetrical/
#48. Lateral leaf-blade lobe <lobation>/
1. entire/
2. pinnatifid/
#49. Lateral leaf-blade lobes <shape in outline>/
1. narrowly elliptic in outline <length to width ratio: 2.5-9.0>/
2. elliptic in outline <length to width ratio: 1.35-2.5 >/
3. broadly elliptic in outline <length to width ratio: 1.0-1.35>/
4. transversely elliptic in outline <length to width ratio: 0.4-0.75>/
5. linear in outline <length to width ratio: greater than 9.0>/
6. narrowly oblong in outline <length to width ratio: 2.5-9.0>/
7. broadly oblong in outline <length to width ratio: 1.0-1.35>/
8. square in outline <length to width ratio: 1.0>/
9. ovate in outline <length to width ratio: 1.35-2.5>/
10. broadly ovate in outline <length to width ratio: 1.0-1.35>/
11. very broadly ovate in outline <length to width ratio: 0.75-1.0>/
12. shallowly ovate in outline <length to width ratio: 0.4-0.75>/
13. very shallowly ovate in outline <length to width ratio: 0.11-0.4>/
14. broadly triangular in outline <length to width ratio: 1.0-1.35>/
15. shallowly triangular in outline <length to width ratio: 0.4-0.75>/
#50. Lateral leaf-blade lobes <shape at apex>/
1. narrowly acute at the apex/
2. acute at the apex/
3. broadly acute at the apex/
4. broadly acuminate at the apex/
5. obtuse at the apex/
#51. Lateral leaf-blade lobes <length in cm>/
cm/
#52. Lateral leaf-blade lobes <width in cm>/
cm/
#53. Tendrils <presence or absence>/
1. present/
2. absent/
#54. Tendrils <number>/
1. solitary/
2. 5-8 at a node in an axillary fascicle/
#55. Tendrils <simple or bifid>/
1. simple/
2. bifid/
#56. Tendrils <length in cm>/
cm long/

#57. Tendrils <aculeate or not>/
      1. not aculeate/
      2. aculeate/
#58. Tendrils <pubescence type>/
      1. glabrous/
      2. glabrate/
      3. hispid/
      4. hispidulous/
      5. hirsute/
      6. pilose/
      7. antrorse-strigose/
      8. retrorse-strigose/
      9. setose/
      10. scabrous/
#59. Inflorescences <sexuality and flower arrangement>/
      1. unisexual <only male or female flowers>/
      2. androgynous <female flowers below male flowers>/
      3. gynecandrous <female flowers above male flowers>/
#60. Bisexual inflorescences <type>/
      1. racemose/
      2. fasciculate/
#60. Bisexual inflorescences <number of flowers>/
      -flowered/
#61. Bisexual inflorescence <peduncle presence or absence>/
      1. pedunculate/
      2. sessile/
#62. Male inflorescences <type>/
      1. modified compound dichasial/
      2. paniculate/
      3. racemose/
      4. fasciculate/
      5. a solitary flower/
#63. Male inflorescences <number of flowers>/
      -flowered/
#64. Male inflorescences <peduncle presence or absence>/
      1. pedunculate/
      2. sessile/
#65. Peduncle <male inflorescence - length in cm>/
      cm long/
#66. Pedicel <male flowers - shape in outline>/
      1. terete in cross section <smooth circular in outline, cylindric>/
      2. sulcate in cross section <alternating longitudinal grooves and
          furrows>/
#67. Pedicel <male flowers - length in mm>/
      mm long/
#68. Pedicel <male flowers - pubescence type>/
      1. glabrous/
      2. glabrate/
      3. hispid/
      4. hispidulous/
      5. hirsute/
      6. pilose/
      7. antrorse-strigose/
      8. retrorse-strigose/
      9. scabrous/
#69. Pedicel <male flowers - presence/absence of bracteole>/
      1. without bracteoles/
      2. subtended by a bracteole/
#70. Bracteole subtending pedicel <male flowers - shape>/
      1. very narrowly elliptic in outline <length to width ratio: 2.5-9.0>/
      2. narrowly long cuneate in outline <length to width ratio: 2.5-9.0>/
#71. Bracteole subtending pedicel <male flowers - shape at base>/
      1. narrowly long cuneate at the base/
      2. FILLER/

#72. Bracteole subtending pedicel <male flowers - shape at apex>/
      1. narrowly acute at the apex/
      2. acute at the apex/
#73. Bracteole subtending pedicel <male flowers - length in mm>/
      mm/
#74. Bracteole subtending pedicel <male flowers - wide in mm>/
      mm/
#75. Bracteole subtending pedicel <male flowers - pubescence type>/
      1. antrorse-strigose at apex; glabrous at base/
      2. FILLER/
#76. Hypanthium <male flowers - shape>/
      1. infundibular/
      2. campanulate/
#77. Hypanthium <male flowers - height in mm>/
      mm long/
#78. Hypanthium <male flowers - diameter in mm>/
      mm in diam/
#79. Hypanthium <male flowers - pubescence type>/
      1. glabrate/
      2. hispid/
      3. hispidulous/
      4. hirsute/
      5. pilose/
      6. antrorse-strigose/
      7. scabrous/
#80. Calyx lobes <male flowers - shape>/
      1. linear in outline <length to width ratio: greater than 9.0>/
      2. narrowly oblong in outline <length to width ratio: 2.5-9.0>/
      3. oblong in outline <length to width ratio: 1.35-2.5>/
      4. narrowly triangular in outline <length to width ratio: 2.5-9.0>/
      5. triangular in outline <length to width ratio: 1.35-2.5>/
      6. broadly triangular in outline <length to width ratio: 1.0-1.35>/
#81. Calyx lobes <male flowers - shape at apex>/
      1. narrowly acute at the apex/
      2. acute at the apex/
#82. Calyx lobes <male flowers - length in mm>/
      mm/
#83. Calyx lobes <male flowers - width in mm>/
      mm/
#84. Calyx lobes <male flowers - pubescence type>/
      1. glabrous/
      2. glabrate/
      3. hispid/
      4. hispidulous/
      5. hirsute/
      6. pilose/
      7. puberulent/
      8. antrorse-strigose/
      9. scabrous/
#85. Corolla <male flowers   overall shape>/
      1. infundibular/
      2. campanulate/
#86. Corolla <male flowers - pubescence type outside>/
      1. glabrous outside/
      2. hispid outside/
      3. hispidulous outside/
      4. hirsute outside/
      5. pilose outside/
      6. puberulent outside/
      7. antrorse-strigose outside/
      8. scabrous outside/
#87. Corolla <male flowers - pubescence type inside>/
      1. glabrous inside/
      2. puberulent inside/
#88. Corolla tube <male flowers - height in mm>/
      mm long/

#89. Corolla tube < male flowers - diameter in mm > /
    mm in diam/
#90. Corolla tube < male flowers - external pubescence type > /
    1. glabrous outside/
    2. hispid outside/
    3. hispidulous outside/
    4. hirsute outside/
    5. pilose outside/
    6. puberulent outside/
    7. antrorse-strigose outside/
#91. Corolla lobes < male flowers - shape > /
    1. elliptic in outline < length to width ratio: 1.35-2.5 > /
    2. broadly elliptic in outline < length to width ratio: 1.0-1.35 > /
    3. oblong in outline < length to width ratio: 1.35-2.5 > /
    4. ovate in outline < length to width ratio: 1.35-2.5 > /
    5. broadly ovate in outline < length to width ratio: 1.0-1.35 > /
    6. very broadly ovate in outline < length to width ratio: 0.75-1.0 > /
    7. shallowly ovate in outline < length to width ratio: 0.4-0.75 > /
    8. narrowly obovate in outline < length to width ratio: 2.5-9.0 > /
    9. obovate in outline < length to width ratio: 1.35-2.5 > /
    10. broadly triangular in outline < length to width ratio: 1.0-1.35 > /
#92. Corolla lobes < male flowers - shape at apex > /
    1. narrowly acute at the apex/
    2. acute at the apex/
    3. broadly acute at the apex/
    4. mucronate at the apex/
    5. obtuse at the apex/
#93. Corolla lobes < male flowers - length in mm > /
    mm/
#94. Corolla lobes < male flowers - width in mm > /
    mm/
#95. Corolla lobes < male flowers - external pubescence type > /
    1. glabrous outside/
    2. glabrate outside/
    3. hispid outside/
    4. hispidulous outside/
    5. hirsute outside/
    6. pilose outside/
    7. puberulent outside/
    8. antrorse-strigose outside/
    9. scabrous outside/
#96. Corolla lobes < male flowers - internal pubescence type > /
    1. glabrous inside/
    2. puberulent inside/

97. Stamens separating from the hypanthium < male flowers - distance from
        the base of the hypanthium in mm > /
    mm from the base of the hypanthium/
#98. Filaments < male flowers - shape in outline > /
    1. terete in cross section < smooth circular in outline, cylindric > /
    2. radially compressed in cross section/
#99. Filaments < male flowers - length in mm > /
    mm long/
#100. Filaments < male flowers - width or diameter in mm > /
    mm wide/
#101. Filaments < male flowers - pubescence type > /
    1. glabrous/
    2. with basal 1/2-2/3 puberulent and glabrous apically/
#102. Anther thecae < male flowers - length in mm > /
    mm long/
#103. Anther thecae < male flowers - pubescence type > /
    1. glabrous/
    2. hispidulous/
    3. puberulous/
    4. puberulent/
    5. papillose/

#104. Anther connective <male flowers - shape>/
    1. elliptic <length to width ratio: 1.35-2.5>/
    2. narrowly oblong <length to width ratio: 2.5-9.0>/
    3. oblong <length to width ratio: 1.35-2.5>/
    4. transversely broadly oblong <length to width ratio: 0.75-1.0>/
    5. ovate <length to width ratio: 1.35-2.5>/
    6. obovate <length to width ratio: 1.35-2.5>/
#105. Anther connective <male flowers - entire or divided at apex>/
    1. unilobate/
    2. bilobate/
#106. Anther connective <male flowers - shape of apex>/
    1. acute at apex/
    2. obtuse at apex/
#107. Anther connective <male flowers - condition of apex>/
    1. smooth at the apex/
    2. puberulous at the apex/
    3. papillate at the apex/
    4. fimbriate at the apex/
    5. crenulate at the apex/
#108. Anther connective <male flowers - length in mm>/
    mm/
#109. Anther connective <male flowers - width in mm>/
    mm/
#110. Disc <male flowers - shape>/
    1. cylindrical/
    2. consisting of 3 papillae/
#111. Disc <male flowers - height in mm>/
    mm long/
#112. Disc <male flowers - diameter in mm>/
    mm in diam/
#113. Female inflorescences <type>/
    1. a solitary flower/
    2. fascicles/
#114. Female inflorescences <number of flowers>/
    -flowered/
#115. Peduncle <female inflorescence - length in cm>/
    cm long/
#116. Pedicel <female flower - shape in outline>/
    1. terete in outline <smooth circular in outline, cylindric>/
    2. sulcate in outline <alternating longitudinal grooves and furrows>/
#117. Pedicel <female flower - length in mm>/
    mm long/
#118. Pedicel <female flower - pubescence type>/
    1. glabrate/
    2. hispid/
    3. hispidulous/
    4. hirsute/
    5. pilose/
    6. antrorse-strigose/
    7. retrorse-strigose/
    8. setose/
    9. scabrous/
#119. Pedicel <female flower - individual hair type>/
    1. with nonbreakaway hairs/
    2. with breakaway hairs <nonglandular, multicellular, conical hairs
        consisting of an easily ruptured, multicellular foot with
        numerous, small, thin-walled cells and a uniseriate body with
        larger, thick-walled cells (Inamdar & Gangadhara, 1975; Inamdar
        et al., 1990)>/
#120. Pedicel <female flower - overall shape>/
    1. cylindrical <the sides parallel from base to apex>/
    2. flaring upwards from a narrower base to a wider apex/
#121. The lower <female flower - amount of hypanthium fused to ovary>/
    1. 2/3 of hypanthium fused to the ovary/
    2. 3/4 of hypanthium fused to the ovary/
    3. 4/5 of hypanthium fused to the ovary/

#122. Fused portion of hypanthium <female flower - shape>/
    1. ellipsoid/
    2. cylindrical/
    3. ovoid/
#123. Fused portion of hypanthium <female flower - height in mm>/
    mm long/
#124. Fused portion of hypanthium <female flower - diameter in mm>/
    mm in diam/
#125. Fused portion of hypanthium <female flower - aculeate or not>/
    1. not aculeate/
    2. aculeate/
#126. Fused portion of hypanthium <female flower - pubescence type>/
    1. glabrous/
    2. hirsute/
    3. pilose/
    4. lanate/
    5. puberulent/
    6. antrorse sericeous/
    7. retrorse sericeous/
    8. scabrous/
#127. Aculei <female flower, lower portion of hypanthium - texture>/
    1. soft/
    2. stiff//
#128. Aculei <female flower, lower portion of hypanthium - length in mm,
        DEPENDENT on 130,1-3/5-7>/
    mm long/
#129. Hyaline bristle at apex of aculeus <female flower, lower portion of
        hypanthium - length in mm>/
    mm long/
#130. Hyaline bristle at apex of aculeus <female flower, lower portion of
        hypanthium - relative length in comparison to opaque base,
        DEPENDENT on 130,1-3/5-7>/
    times as long as the opaque base/
#131. The upper <female flower - amount of hypanthium free from ovary>/
    1. 1/3 of hypanthium free from the ovary/
    2. 1/4 of hypanthium free from the ovary/
    3. 1/5 of hypanthium free from the ovary/
#132. Free portion of hypanthium <female flower - height in mm>/
    mm long/
#133. Free portion of hypanthium <female flower - diameter in mm>/
    mm in diam/
#134. Free portion of hypanthium <female flower - aculeate or not>/
    1. not aculeate outside/
    2. aculeate outside/
#135. Free portion of hypanthium <female flower - pubescence type outside>/
    1. glabrous outside/
    2. hispid outside/
    3. hispidulous outside/
    4. hirsute outside/
    5. pilose outside/
    6. puberulent outside/
    7. antrorse-strigose outside/
#136. Free portion of hypanthium <female flower - pubescence type inside>/
    1. glabrous inside/
    2. puberulent inside/
#137. Calyx lobes <female flower - shape>/
    1. narrowly elliptic in outline <length to width ratio: 2.5-9.0>/
    2. linear in outline <length to width ratio: greater than 9.0>/
    3. narrowly oblong in outline <length to width ratio: 2.5-9.0>/
    4. narrowly obovate in outline <length to width ratio: 2.5-9.0>/
    5. narrowly triangular in outline <length to width ratio: 2.5-9.0>/
    6. triangular in outline <length to width ratio: 1.35-2.5>/
#138. Calyx lobes <female flower - shape at apex>/
    1. narrowly acute at the apex/
    2. acute at the apex/

#139. Calyx lobes <female flower - length in mm>/
    mm/
#140. Calyx lobes <female flower - width in mm>/
    mm/
#141. Calyx lobes <female flower - pubescence type>/
    1. glabrate/
    2. hispid/
    3. hispidulous/
    4. hirsute/
    5. puberulent/
    6. pilose/
    7. antrorse-strigose/
    8. scabrous/
#142. Corolla <female flower - pubescence type outside>/
    1. hispid outside/
    2. hispidulous outside/
    3. hirsute outside/
    4. pilose outside/
    5. puberulent outside/
    6. antrorse-strigose outside/
#143. Corolla <female flower - pubescence type inside>/
    1. glabrous inside/
    2. puberulent inside/
#144. Corolla tube <female flower - presence>/
    1. present/
    2. absent/
#145. Corolla tube <female flower - height in mm>/
    mm long/
#146. Corolla tube <female flower - diameter in mm>/
    mm in diam/
#147. Corolla tube <female flowers - external pubescence type>/
    1. glabrous outside/
    2. hispid outside/
    3. hispidulous outside/
    4. hirsute outside/
    5. pilose outside/
    6. puberulent outside/
    7. antrorse-strigose outside/
#148. Corolla lobes <female flower - shape>/
    1. elliptic in outline <length to width ratio: 1.35-2.5>/
    2. broadly elliptic in outline <length to width ratio: 1.0-1.35>/
    3. oblong in outline <length to width ratio: 1.35-2.5>/
    4. ovate in outline <length to width ratio: 1.35-2.5>/
    5. broadly ovate in outline <length to width ratio: 1.0-1.35>/
    6. obovate in outline <length to width ratio: 1.35-2.5>/
    7. broadly obovate in outline <length to width ratio: 1.0-1.35>/
#149. Corolla lobes <female flower - shape at apex>/
    1. broadly acute at the apex/
    2. acute at the apex/
    3. mucronate at the apex/
    4. obtuse at the apex/
#150. Corolla lobes <female flower - length in mm>/
    mm/
#151. Corolla lobes <female flower - width in mm>/
    mm/
#152. Corolla lobes <female flowers - external pubescence type>/
    1. glabrous outside/
    2. glabrate outside/
    3. hispid outside/
    4. hispidulous outside/
    5. hirsute outside/
    6. pilose outside/
    7. puberulent outside/
    8. antrorse-strigose outside/
#153. Corolla lobes <female flowers - internal pubescence type>/
    1. glabrous inside/

    2. hispidulous inside/
    3. puberulent inside/
#154. Staminodes <female flower - presence/absence>/
    1. present/
    2. absent/
#155. Staminodes separating from the free portion of the hypanthium <female
        flower - distance above the ovary on the free portion of the
        hypanthium in mm>/
    mm above the ovary/
#156. Staminodes <female flower - height in mm>/
    mm long/
#157. Staminodes <female flower - diameter in mm>/
    mm in diam/
#158. Staminodes <female flower - pubescence type>/
    1. glabrous/
    2. puberulent/
#159. Style <female flower - height in mm>/
    mm long/
#160. Style <female flower - diameter in mm>/
    mm in diam/
#161. Style <female flower - disc presence>/
    1. subtended by a circular disc/
    2. not subtended by a disc/
#162. Disc <female flowers - height in mm>/
    mm long/
#163. Disc <female flowers - diameter in mm>/
    mm in diam/
#164. Stigma <female flower - height in mm>/
    mm long/
#165. Stigma <female flower - diameter in mm>/
    mm in diam/
#166. Stigma <female flower - lobation>/
    1. entire/
    2. sublobate/
    3. lobate/
#167. Stigma <female flower - number of lobes>/
    1. with 3 finger-like projections on the margin/
    2. with 4 finger-like projections on the margin/
    3. with 5 finger-like projections on the margin/
    4. with 6 finger-like projections on the margin/
    5. with 9 finger-like projections on the margin/
#168. Finger-like stigmatic projections <female flowers - height in mm,
        DEPENDENT on 177,1>/
    mm long/
#169. Finger-like stigmatic projections <female flowers - diam in mm,
        DEPENDENT on 177,1>/
    mm in diam/
#170. Finger-like stigmatic projections <female flowers - surface>/
    1. smooth/
    2. papillate/
#171. Fruit <nongeocarpic or geocarpic>/
    1. not geocarpic, maturing above ground and readily visible/
    2. geocarpic, maturing below ground and not visible/
#172. Pedicel <fruit - shape in transverse section>/
    1. terete in cross section <smooth circular in outline, cylindric>/
    2. sulcate in cross section <alternating longitudinal grooves and
        furrows>/
#173. Pedicel <fruit - length in cm>/
    cm long/
#174. Pedicel <fruit - pubescence type>/
    1. glabrous/
    2. glabrate/
    3. hispid/
    4. hispidulous/
    5. hirsute/
    6. pilose/

       7. antrorse-strigose/
       8. retrorse-strigose/
       9. setose/
#175. Pedicel < fruit - individual hair type > /
       1. with nonbreakaway hairs/
       2. with breakaway hairs < nonglandular, multicellular, conical hairs
          consisting of an easily ruptured, multicellular foot with numerous,
          small, thin-walled cells and a uniseriate body with larger,
          thick-walled cells (Inamdar & Gangadhara, 1975; Inamdar et al.,
          1990) > /
#176. Pedicel < fruit - overall shape > /
       1. cylindrical < the sides parallel from base to apex > /
       2. flaring upwards from a narrower base to a wider apex/
#177. Fruit < coloration - single or bicolored > /
       1. monocolored/
       2. bicolored with longitudinal stripes from base to apex/
#178. Fruit < coloration - basal color > /
       1. green/
       2. red/
       3. yellow/
       4. white/
       5. brown/
       6. orange/
#179. Fruit < coloration - longitudinal stripes color > /
       1. with light green longitudinal stripes/
       2. with dark green longitudinal stripes/
       3. with yellow longitudinal stripes/
       4. with brown longitudinal stripes/
       5. with white longitudinal stripes/
       6. with purple longitudinal stripes/
       7. with orange longitudinal stripes/
#180. Fruit < shape > /
       1. ellipsoid < height to diameter ratio: 1.35-2.5 > /
       2. globose < height to diameter ratio: 1.0 > /
       3. cylindrical < height to diameter ratio: 1.35-2.5 > /
       4. ovoid < height to diameter ratio: 1.35-2.5 > /
       5. obovoid < height to diameter ratio: 1.35-2.5 > /
       6. spindle-shaped < height to diameter ratio: 1.35-2.5 > /
#181. Fruit < length in cm > /
       cm long/
#182. Fruit < diameter in cm > /
       cm in diam/
#183. Fruit < surface - aculeate or not > /
       1. not aculeate/
       2. aculeate/
#184. Fruit < surface - pubescence type > /
       1. glabrous/
       2. hispidulous/
       3. hirsute/
       4. pilose/
       5. puberulent/
       6. retrorse-strigose/
#185. Fruit < apex shape > /
       1. blunt at the apex < without a beak at the apex IMPLICIT > /
       2. rostrate at the apex/
#186. Beak < length in cm DEPENDENT on 220,1 > /
       cm long/
#187. Aculei < fruit - shape > /
       1. terete < smooth circular in outline, cylindric > /
       2. laterally compressed near the base/
#188. Aculei < fruit - hyaline bristle present or absent > /
       1. with hyaline bristle persistent/
       2. with hyaline bristle lost/
#189. Aculei < fruit - height in mm > /
       mm long/

#190. Aculei <fruit - diameter in mm>/
    mm in diam/
#191. Bristle <aculei of fruit - length in mm>/
    mm long/
#192. Seeds <shape>/
    1. elliptic <length to width ratio: 1.35-2.5>/
    2. ovate <length to width ratio: 1.35-2.5>/
#193. Seeds <length in mm>/
    mm/
#194. Seeds <width in mm>/
    mm/
#195. Seeds <thickness in mm>/
    mm thick/
#196. Seeds <presence of apical wing>/
    1. unwinged/
    2. apically winged/
#197. Wing <shape>/
    1. transversely oblong <length to width ratio: 0.4-0.75>/
    2. FILLER/
#198. Wing <shape of apex>/
    1. truncate at apex <DEPENDENT on 204,1>/
    2. FILLER/
#199. Wing <length in mm>/
    mm/
#200. Wing <width in mm>/
    mm/
#203. 2N = <diploid chromosome number>/
    /
#204. Native continental distribution:/
    1. Africa/
    2. Southwest Asia <character 193, Middle East>/
    3. Asia/
    4. Malesia <sensu *Flora Malesiana*/
    5. Australia/
#205. Africa: <native distribution>/
    1. Angola <formerly Portuguese West Africa>/
    2. Benin <formerly Dahomey>/
    3. Botswana <formerly Bechuanaland>/
    4. Burkina <formerly Upper Volta>/
    5. Burundi <formerly Urundi>/
    6. Cameroon/
    7. Cape Verde Islands/
    8. Central African Republic <formerly Ubangi-Shari, Oubangi-
        Chari, or Central African Empire>/
    9. Chad <or Tchad>/
    10. Congo <formerly Middle Congo>/
    11. Egypt/
    12. Ethiopia <or Abyssinia>/
    13. Gambia/
    14. Ghana/
    15. Guinea-Bissau/
    16. Ivory Coast/
    17. Kenya <formerly East Africa Protectorate>/
    18. Lesotho <formerly Basutoland>/
    19. Liberia/
    20. Madagascar <now Malagasy Republic>/
    21. Malawi <formerly Nyasaland or British Central Africa
        Protectorate>/
    22. Maldive Islands/
    23. Mali <formerly Sudanese Republic or French Sudan>/
    24. Mauritania/
    25. Mozambique <formerly Portuguese East Africa>/
    26. Namibia <formerly South-West Africa>/
    27. Niger/
    28. Nigeria/
    29. Rwanda <formerly Ruanda>/

    30. Senegal/
    31. Seychelles/
    32. Sierra Leone/
    33. Somalia/
    34. South Africa/
    35. Sudan <formerly Anglo-Egyptian Sudan>/
    36. Swaziland/
    37. Tanzania <formerly German East Africa or Tanganyika>/
    38. Uganda/
    39. Zaire <formerly Congo, Belgian Congo, or Congo Free State>/
    40. Zambia <formerly Northern Rhodesia>/
    41. Zimbabwe <formerly Rhodesia>/
#206. Southwest Asia: <native distribution, Middle East>/
    1. Iran/
    2. Iraq/
    3. Israel/
    4. Jordan/
    5. Oman/
    6. Qatar/
    7. Saudi Arabia/
    8. Socotra/
    9. South Yemen/
    10. Syria/
    11. United Arab Emirates/
    12. Yemen/
#207. Asia: <native distribution>/
    1. Afghanistan/
    2. Bangladesh/
    3. Bhutan/
    4. Burma/
    5. Cambodia/
    6. China/
    7. India/
    8. Japan/
    9. Korea/
    10. Laos/
    11. Nepal/
    12. Pakistan/
    13. Sri Lanka/
    14. Taiwan/
    15. Thailand/
    16. Vietnam/
#208. Malesia: <native distribution, sensu *Flora Malesiana*>/
    1. Indonesia/
    2. New Guinea/
    3. Philippines/
#209. Australian: <native distribution, Australia>/
    1. Australia/
    2. FILLER/
#212. Pacific: <native distribution>/
    1. Guam/
    2. Fiji Islands/
    3. New Britain/
    4. Papua New Guinea/
    5. Samoa/
    6. Solomon Islands/
    7. Tonga Islands/
#211. Introduced: <distribution - introduced>/
    1. Cape Verde Islands/
    2. widely cultivated in tropical areas/
    3. widely cultivated all over the world/

Appendix 4. Numerical List of Taxa

1.   Cucumis myriocarpus
1a. Cucumis myriocarpus subsp. myriocarpus
1b. Cucumis myriocarpus subsp. leptodermis
2.   Cucumis africanus
3.   Cucumis quintanilhae
4.   Cucumis heptadactylus
5.   Cucumis kalahariensis
6.   Cucumis anguria
6a. Cucumis anguria var. anguria
6b. Cucumis anguria var. longaculeatus
7.   Cucumis sacleuxii
8.   Cucumis carolinus
9.   Cucumis dipsaceus
10. Cucumis prophetarum
10a. Cucumis prophetarum subsp. prophetarum
10b. Cucumis prophetarum subsp. dissectus
11. Cucumis pubituberculatus
12. Cucumis zeyheri
13. Cucumis prolatior
14. Cucumis insignis
15. Cucumis globosus
16. Cucumis thulinianus
17. Cucumis ficifolius
18. Cucumis aculeatus
19. Cucumis pustulatus
20. Cucumis meeusei
21. Cucumis jeffreyanus
22. Cucumis hastatus
23. Cucumis rigidus
24. Cucumis baladensis
25. Cucumis metuliferus
26. Cucumis rostratus
27. Cucumis hirsutus
28. Cucumis humifructus
29. Cucumis melo
29a. Cucumis melo subsp. melo
29b. Cucumis melo subsp. agrestis
30. Cucumis sagittatus
31. Cucumis sativus
32. Cucumis hystrix

INDEXES

1. Index to Exsiccatae

The numbers in parenthesis refer to the corresponding taxa in the text and in Appendix 4, Numerical List of Taxa.

Abeleven sn (29).
Ables 1988 (31).
Ackland 111(1).
Acocks 12248(27); 12606(2); 14382(30); 1538(4); 1579(4); 16613(27); 18028(20); 1856(4); 18693(14); 18821(2); 18831(20); 19178(14); 19218(14); 19575(14); 21678(14); 21960(14); 5753(30).
Acocks & Hafström H1389(2).
J.G. Adam 13020(10b); 13149(10b); 19218(10b); 19283(29b); 19289(10b); 21729(29b); 21799-21(10b).
C.D. Adams 3463(29a).
C.D. Adams & Shafeeg 14934(29b).
J. Adamson 63(10b).
R.S. Adamson 3134(1b).
Aedegeb 5 Aug 1924 (31).
Afzelius sn (29).
Agnew 8204(18); 9782(9).
Aguiar Macêdo 4741(6b); 5037(6b); 5506(27).
Ahrens Jul 1950 (2).
Ainsbury 3076(29).
Aitchison 173(29a); 192(29b); 196(29); 464(29a); 983(29a).
Ake-Assi 5495(26).
Al-Rawi 19917(29a).
d' Alleizette Aug 1909 (9).
J.B. Allen & Elmi 546(10b).
A. Allen 285(2).
Alston 2116(29).
St. Eloy D' Alton 79(1b).
Amdjah 496(29b).
Ancher 1833 (10a).
Anchieta 18(30).
A. Anderson XLI(31).
B. Anderson 1149(6b); 1878(29b).
J.R. Anderson ORFS276-b(4).
P. Anderson A107(2).
T. Anderson May 1859 (10a).
Andorfer sn (31).
Andrew 277(29).
Andrews 138(29); 183(29); 218(29).

F.W. Andrews A79(29b).
Angus 1367(6b).
Antonio & D'Arcy 2469(29).
Aplin 6279(29a).
Araquistain 326(6a).
Arboleda, Bunch & Von Loh 169(9).
Arceneaux 120(29b).
Archbold 1378(7).
Archibald 4156(14).
Armit 13(29).
Armstrong sn (29b).
Arvidsson 84(10b).
Ashall CA 4(8).
Ashbey 22(29).
Ash 1570(10b); 2458(19); 2502(29b); 2505(21); 263(17); 2776(17).
Asplund 15267(9).
Aston 2194(1b).
Atasrip 192(29); 85(29).
Aublet 20(29b).
Aucher-Eloy sn (29a); 2857(10a); 4503(10b).
Augustinowiz 20 Apr 1872 (31).
Axtell & Axtell 2898(31).
Axtell, Axtell & Axtell 2899(29).
Aylamer 597(10a).
Azancot de Menezes 1310(27); 2424(27).
Azancot de Menezes & Henriques 205(30).
Backer 3720(29b).
Backéus 1099(9); 1172(25); 1191(25).
Baijnath 159(6b).
D.M. Bailey Jun 1892 (29a).
L.H. Bailey 10(9); 211(9); 1891 (31); 1897 (9); 3031(31); 4980(6a); 62150(6a); 6980(29a); 13439(31); 13819(31); 13839(31); 14039(31); 14079(31); 14119(31); 14159(31); 14239(31); 14259(6a); 30109(29a); 30509(29); 30529(29a); 30629(29); 58100(29a); 4 Aug 1885 (31); 15 Aug 1885 (29a); 1 Sep 1887 (6a); 4 Sep 1889 (29b); 9 Sep 1889 (29a); 11 Sep 1889 (29a); 11 Sep 1889 (31); 12 Sep 1889 (31); 13 Sep 1889 (29a); 13 Sep 1889 (31); 15 Sep 1889 (29a); 18 Sep 1889 (29a); 18 Sep 1889 (31); 21 Apr 1890 (31); 14 May 1890

(31); 16 Jun 1890 (31); 27 Aug 1890 (31); 30 Aug
1890 (31); 1 Sep 1890 (31); 9 Sep 1890 (31); 13
Sep 1890 (29); 14 Sep 1890 (9); 14 Sep 1890
(29b); 15 Sep 1890 (6a); 15 Sep 1890 (9); 15 Sep
1890 (29a); 16 Sep 1890 (29); 20 Sep 1890 (31);
27 Sep 1890 (31); 11 Oct 1890 (29); 28 Mar 1891
(31); 31 Aug 1891 (29a); 18 Sep 1891 (29a); 1 Oct
1891 (29a); 3 Oct 1891 (31); 24 Aug 1892 (31);
Sep 1892 (31); Feb 1893 (31); 26 Sep 1893 (31);
14 Feb 1894 (29a); Feb 1894 (29a); 26 Sep 1894
(29); 13 Sep 1895 (29a); 15 Sep 1895 (29); 8 Nov
1895 (31); 29 Jan 1896 (29a); 5 Sep 1917 (31); 14
Sep 1917 (29a); 15 Sep 1917 (29); 15 Sep 1917
(31); 22 Sep 1917 (29a); 22 Sep 1917 (31); 28 Sep
1917 (29); 10 Aug 1918 (31); 26 Aug 1918 (31);
22 Aug 1919 (29a); 10 Sep 1919 (29); 5 Aug 1920
(29); 22 Aug 1920 (6a); 18 Sep 1920 (29a); 20 Sep
1920 (29a); Sep 1920 (29a); Sep 1920 (31); 1 Oct
1920 (29a); 1 Oct 1920 (31); 2 Oct 1920 (31); 4
Oct 1920 (31); 27 Jul 1923 (31); 13 Aug 1923 (31);
16 Aug 1924 (31); 19 Sep 1925 (29); 1 Aug 1927
(29); 10 Sep 1927 (29a); 11 Sep 1927 (29a); 3 Jul
1934 (25); 18 Aug 1936 (9); 18 Aug 1936 (29a); 9
Sep 1936 (29a); 4 Jul 1938 (29); 5 Jul 1938 (29);
22 Aug 1938 (29); 24 Aug 1938 (29); 28 Aug 1938
(29); 3 Sep 1938 (9); 3 Sep 1938 (29a); 23 Sep
1938 (2); 23 Sep 1938 (9); 27 Sep 1938 (29); 19
Sep 1938 (29a); 2 Jul 1939 (29); 8 Sep 1940 (6a); 1
Aug 1943 (9); 15 Aug 1943 (9); 26 May 1944
(29b); 13 Aug 1944 (29); 20 Aug 1944 (29b); 9
Sep 1944 (29); 1 Sep 1945 (31); 27 Jul 1947 (29).
L.H. Bailey & E.Z. Bailey 1042(6a); 1568(29b).
Bakanou 9743(29b).
Baker & O'Donovan 3992(6a).
A. Baker 23(1a).
C.F. Baker 20(29a).
Bakhuizen van der Brink 7619(29b).
Baldwin, J 13591(29a).
Balfour 52(10b); 682(10b); 1880 (10a).
Ballance 613(2).
Ballin sn (6a).
Bally 9979(10b); B3645(13); B6787(10b); B6955(10a).
Bally & Melville 15326(10b); 16028(16); 16148(10b).
Balsinhas 1346(2); 1349(27); 1468(25); 2971(27);
3142(6b); 3354(27); 3384(14); 3621(29); 3625(27).
Bamer 57/62(29a).
Bamps 6695(10b); 7616(10a); 7622(29); 7684(29b).
Bamps & Martins 4326(25).
Baneckurt 30(1a).
Banerjee 115(29).
Banks & Solander 1769 (29a); sn (29).
Barber 189(29b); 2112(31).
Barbosa 652(2); 5758(6a).
Barbosa & Correia 9101(2); 9103(30); 9114(30).
Barbosa & de Lemos 8592(25).
Barbosa & Magalhães 14459(6b).
Barbosa & Moreno 9862(27); 10202(6b).
Barbosa & Silva 14086(6b); 14375(6b).
Barkley 13372(29).
W.G. Barnard & Mogg 629(6b).
D.E. Barnard CUR 2(29).
Baron 2345(7); 6562(7).
Barranca 485(29).
Barrett 497(1a).
Barros Machado 119(27); 311(27).

Barter 1527(25).
Bartha 18(29b); 18/26(29); 58(10b); 58/27(10a); Jul
1967 (29a).
Basedow 29(1b).
Basera 268(14).
Batanouny 1271(10a).
Bates 93976(1a).
Batty 157(7); 1141(19).
Bauml 29-336(29).
Baur 292(27).
Bax 62(10b); 63(19); 1932 (27).
Bayliss 8483(2).
Baytop 14405(29a); 14406(29a).
Beals 72(10a).
Beard 8489(29).
Beauglehole ACB 24926(1b); ACB 3121(1b); ACB
55767(1); ACB 55856(1b); ACB 63944(1); ACB
67696A(1b); ACB 67927(1); ACB 68802(1b).
Beauglehole & Finck ACB 29593(1b); ACB 33116(1).
Beauglehole & Errey ACB 70451(1b).
Beauglehole & Macfarlane ACB 55611(1b).
Becker 11418(1b).
Beckett 1810(10a).
Becquet 1040(2); 5252(17).
Behr 238(14).
Belcher 893(31).
Bell 326(29).
Belo & Marques 43(27).
Bent 1895 (10a); 1896 (10a).
Bent & Bent 1897 (10b).
Benzon sn (31).
Berg 1843 (1).
Berger 369(1b).
W. Berger 2883(29b).
Bergler sn (31).
Berhaut 534(29); 1196(25); 1612(10a); 5604(10a);
7478(25).
van Berkel 87(23).
Beskin 91(29a).
Best 666(27).
Betche Feb 306(29b); 1893 (1b).
Beverly & Hoener 537(1b).
Bhandani 50(10a).
Bianor 528(29a).
Biegel 751(1a); 1490(14); 2183(25); 2557(27);
2864(27).
Biegel, Müller & Gibbs Russell 5077(14).
Bingham 1039(28).
Birschop 2364(29a).
Bishop 2364(31).
Blackmore, Banda & Patel 222(27).
Black 7(1b).
Blake 17842(29b).
Blakely & Shiress 13 Feb 1920 (1b).
Blanchet 1010(6a).
Bloembergen 3042(29).
Blytt 5434(29b).
Bock 41(27).
Bodinier Aug 1888 (29b).
Boerhare sn (29).
Boerlage 27 Oct 1888 (31); 16 Feb 1889 (31).
Boettger 73-70(PI 214050)(29a); 73-91(14); 74-28(PI
282444)(30); 74-35(19); 74-36(29b); 74-46(25);
74-53(PI 374153)(1b); 74-66(29); 73-154(29);
73-155(6b); 73-156(6b); 73-158(PI 140471)(29);

Aug 1973 (2).
Bohnen 7222(1b).
Boissier 1022(10a).
Boivin sn (29b).
Bojko 9 Mar 1936 (10a).
Bole 533(29a); 2212(31).
F. Bolus 39(1b); 6451(27).
H. Bolus 666(1b); 666(1b); 7774(27); 7775(27);
  8913(27); 10386(2); 11896(14); 13109(1b);
  13110(4).
Bondam Oct 1896 (29); Oct 1896 (31).
Bonney sn (29).
Bonny 26(19).
Boorman Feb 1899 (1b); Mar 1900 (29); Mar 1909
  (1b); 30 Jan 1911 (29); Feb 1911 (1b).
Borgesen 14(29a); 146(10a).
Borges 26(6b).
Borggren sn (31).
Borissow Oct 1957 (18).
Bornmüller 358(10b); Aug 1908 (29a).
Boss 55909(20).
Bos 2146(31).
Botha C 14(25).
Boulos 5404(10a); 9285(18).
Boulos & Ads 14197(10a).
Boulos & Getahun 11870(10b).
Boulos, Jallad, Lahham & Abu Hmaidan 7525(10a).
Bourne 3307(29a).
Bourquin 639(14); 822(14).
Bouton 1804 (6a).
Bouxin & Radoux INRS/1072(17); INRS/1208(17);
  INRS/1944(10b).
Bowman 81(29); 390(29a); 1871 (29a).
Boyko 4 Jun 1949 (10a).
Brandegee 16 Oct 1899 (9).
Brass 24287(29b).
Breakwell 1912 (1); Feb 1914 (1).
van Breda 4483(1b).
Bredenkamp 1591(6b).
Bredenkamp & van Vuuren 6(14).
Bredo 2020(29).
Breedlove 20193(6a).
Breito 992(29).
Brenan 8875(26).
Bretschneider 309(29).
Breyer 20659(6b); Jan 1919 (6b).
Brierley 141(14).
Briley & Carroll 1584(29b).
Bristan 5(9); P/12(29b).
Broadway 4553 Ser I(6a).
Brontel sn (1b).
Broun & Broun 1426(19).
Brown & Shapiro 292(6b).
C.A. Brown 9828(29b); 9829(29b); 9850(29b).
R.C. Brown 25 Jan 1970 (2).
Bruce 911(27).
Bruckner 1888 (1).
Brueckner 30(14); 1103(2).
Brumbach 6889(29).
Brummitt 9548(6b).
Brunt 2115(6a).
Bryant Apr 1921 (1b); Apr 1921 (2).
F. Buchanan (Hamilton) 2153A(31); 2153B(31);
  2154(29b); 2154C(29a).
J. Buchanan 100(31); 108(29); 176(27); 16 Feb 1878

(25).
Buitendag 1006(25).
Bullock 2283(17); 2370(27); 3035(19); 18 Feb 1950
  (29b).
Bunbury 1880 (29); 1882 (29).
Bunting 5294(6a); 5981(29b); 9971(29).
Bürger sn (31).
G.H. Burger 546(27).
W. Burger 2883(29b); 3498(17); 3518(10b).
Burke sn (1a); sn (27); Jan (4); Feb (4).
van Burkom sn (29).
Burrows M20(1).
Burtt Davy 374(14); 11773JBD(4).
Bury 31(10a).
Busse 857(27).
Buswell 29 Oct 1948 (29).
Buwalda 1119(31).
Buysman 24(9); 382(29a).
Byrnes 1375(29b); 2585(29a); 2614(29a).
Caley 7 Jun 1956 (9).
Callens 2816(27); 3099(27).
Cambage 3774(1b); 4276(1b).
Camerik 158(1a); 91(27).
J. Campbell 66(29).
J.S. Campbell 22 Feb 1959 (6a).
Capêlo & Ivens 10(30).
Cardosa de Matos 5455(6a).
Cardoso de Matos 5224(6b); 5263(6b); 5291(29a);
  5651(6b).
J.A. Cardoso sn (6a).
Carey 1878 (29).
W. Carne Mar 1914 (29).
W.M. Carne 294(10a).
Carpenter 322(29a).
Carrington 20 Apr 1977 (29a).
Carr 950(29b).
J.H. Carter 947(1a).
S. Carter & Stannard 251(10b); 697(8).
Casey 1879 (29a).
Castro 136(30).
Cerrate 4928(9); 810(9).
Cerrate, Encarnacion & Ferreyra 5159(9).
Chalmers & Bridge Jan 1885 (29).
Chandler 1062(19); 2395(17).
Chandrabose 49155.
Chang 567(29); 568(29b).
Chapard & Villers 22 Aug 1947 (10a).
Chase 728(25); 5331(27); 6729(14); 6832(27);
  7039(25); 8215(27); 8485(14).
Chassignol Aug 1921 (1b).
Cheel 14 Mar 1924 (1b).
Combs 38(6a).
Comins 1002(14); 641(2); 693(1b); 768(14).

Cheesman 10 Aug 1926 (17).
Cheo & Yen 206(29).
Chevalier 1252(10a); 2982(29); 9590(29b); 10463(19).
Chiao 2959(29).
Chippendale 1951(29a).
Chubb 322(27); 328(27).
Chin See Chung 2550(31).
Chun 8382(29b).
Claessens 3(29); 4(29); 854(27).
Clarke 7533(29b); 11510(29); 16823(29); 40777(32).
Cleghorn 1856 (31).

Grontmij (De Bilt) 1(31).
Gross 6 Aug 1912 (29a).
Gueinzius 397(27); 398(14).
Guerra 3811(29).
Guthrie 16044(25).
Gwanisky 165(31).
Gwynne 264(10b).
Haage 3856(29); 3967(29).
Haage & Schmidt 5623(6b); 5624(6b).
Haerdi 381/0(7); 392(27).
Hafström 1389(14).
Hagerup 114(10a); 199(10a); 286(29a); 520(10a).
Haines 2127(31); 2669(29); 2670(29); 420(31); 4423(31).
Hall-Martin 1310(27).
Hamilton Mar 1916 (1b).
Hammel 5260(6a).
Handel-Mazzetti 1972(29); 3111(29).
Hanekom 58a(20); 1335(1a); 1744(1b); 2005(14); 2202(1a); 2203(1a).
Hanid, Hanid & Kiniaruh 538(10b).
Hann 1887 (29).
O.J. Hansen 14(18); 6019(10a).
O.T. Hansen 6018(10b); 6019(10b).
C.C. Harbor 6467(1); 6608(1a).
W.A.H. Harbor 14050(1a).
Hardy 663(1); 1961(20).
Hardy & de Winter 1388(2); 1389(20).
Harland 848(10a); 1319(10a).
Harley 9509(7).
Harlion sn (29).
Harms 30 Jul 1921 (17); 1921 (9); Sep 1923 (31).
Harper 21 Jul 1944 (29a).
Harriman 14640a(6a).
B.J. Harris BJH 480(10b); 2128(9).
B.J. Harris & Harris 4833(18).
W. Harris 6650(6a); 6820(6a); 6967(6a); 9043(6a); 9043(6a).
Hartridge 146900(9).
Hart Nov 1883 (10a); 1883 (29).
Harwood 73(15); 92(15).
Hassib 4 Jul 1933 (29a); 25 Jul 1933 (29a); 20 Jul 1935 (29).
Haussknecht 1867 (29a); 1867 (29a); 21 Jul 1865 (29a); 1865 (29a); Jun 1867 (29a).
Hedenberg 1830 (29); sn (10a).
Hedrén 495(10b).
Hein 27 Mar 1902 (10a).
Helms 539(1b).
Hemming 2011(16).
Hendon 841(29); sn (17).
Henne sn (29a).
Henrici 1569(6b).
Henriques 747(27).
Henry 295(29b); 1876 (29); 8510(29); 12727(32); 13360(32); sn (29).
Hentici 4545(14).
Henty NGF 9840(29); 10593(29b); 14865(29b).
Henty & Lelean NGF 41884(29b).
Hepper 3661(29a).
Hepper & Jaeger 7158(10b).
Hepper & Wood 5999(25).
Hepp 11 Aug 1947 (31).
Herbich sn (31).
Herbst 756(9).

Herbst & Spence 4024(9).
von Herder 4416(29a); 9 Sep 1864 (1b).
Hermann 12 Aug 1833 (31).
Herman 553(1b).
van Heurn Aug 1955 (31).
Hewer 1477(29a); H.4009(29b).
Hewrt 29 Jul 1778 (6a).
Heyne 6735(29); 5 Jan (29a).
Hiemstra 220(14).
Hildebrandt 83a(10a); 150(19); 780(10a); 780b(10a).
Hillcoat 235(10a).
Hilliard & Burtt 3271(14).
Hillier 955(17).
Hitchins & Ward 3(25).
P.M. Hitchins 457(14); 917(14).
Ho-Ch'ang Chow 858(31).
Hobson 240(20).
Hock Feb 1910 (27).
Hodgkin & Stansfield 453(29).
Hofm Aug 1848 (31).
Hohenack 43(29a).
Hokker & Hanbury 1860 (10a).
Holding 1887 (1b).
Holland 220(2).
Holm-Nielsen, Jeppesen, Lojtnant & Ollgaard 2391(9); 2467(9).
Holtze 698(29a).
M. Holtze 409(29a).
W. Holtze 89(29a); 1894 (29a).
Holt 43(9).
Homblé 227(27); 394(31); 1200(27).
Hoogerwerf 15(29).
Hooker sn (31).
Hooper & Gandhi HFP 2462(31).
Hopkins 12(14); 8621(14); 8938(1a); 31 Mar 1942 (25).
A.J.W. Hornby 3588(6a); 4518(27); 4535(6b).
H.I.E. Hornby & R.M. Hornby 718(6b); 774(25).
R.M. Hornby 3343(28).
Horsfield 125(31); sn (29).
Hosaka 2237(9).
Howard 11012(6a).
Hoyle 925(10a).
Hubbard & Winders 6084(1a).
Hucks 72(9).
Hudson 1848 (10a).
J. Hudson 762(9); 947(9).
Huft 1759(29a).
Hughes 4 Dec 1950 (14).
Huliss 5 1904 (1b).
Humbert 16443(2).
Humbles 10010(10a); 10139(10a).
Hunt 191A(1b).
Huq 4314(29).
Hutching 5(1b).
Hutchinson 91(10b); 92(10b).
Hutchinson & Wright 3470(9).
Hu 7702(29).
von Hügel 291(31); 835(31); 1574(31); 1741(31); 1934(29); 2910(29); 3050(29).
Iltis & Iltis 1639(6a).
M Imam 11 Sep 1956 (10a).
Immelamn 589(6b).
Irvine 778(29b); 1819(29a).
Isani NI4635(31); NI5476(31).
Isert 867(6a); sn (29).

Rottler 9 Jan 1800 (29b); 1875 (29a); 1875 (31); sn (29b).
le Roux 660(2); 875(6b).
Roxburgh sn (29b).
Royal Botanic Gardens, Kew 392-603321(29a); 392-603322(29a); 392-603323(29a); 392-603324(29); 392-603325(29a); 392-603326(29a); 602-68(31); 19 Sep 1857 (10b); Aug 1875 (31); 30 Aug 1885 (31); 26 Jul 1910 (25).
Royen sn (29); 323/31(29b); 336/31(29); 406/31(29); 408/31(29a).
Royle 71(31); 80/15(29b); S305(29b); S372 & 376(29); sn (29); sn (29a); sn (31).
Rudatis 1462(27).
Ruffo 1193(9); 1280(19).
Runyinya 226(17).
Runyon 6 Jun 1944 (9).
Rusd 4518(30).
Rushworth 151(6b).
Rutherford-Smith 552(27).
Rwaburindore 212(17).
Sabransky Jul 1910 (31).
de Saeger 1323(27).
Sagot 274(6a).
Sakurai 14 Aug 1910 (29a).
Salada NSW 146903(6a).
Saldanha 14724(31); 15278(31).
Saldanha & Keshava Murthy KFP 3359(31).
Salesiens 1087(6b).
Salter 8605B(1a).
Salubeni 1264(25); 1288(6b).
Salubeni, Banda & Tawkali 2446(27).
Samuelsson 627(10a); 2917(10a); 3153(10a); 4 Apr 1930 (10a); 27 Apr 1932 (10a).
Sanane 546(27).
Sankey sn (27).
Santapau 1056(31); 2565(31); 2918(31); 7635(29b); 7672(10a); 11344(31); 13583(29b); 15181(29b); 16608(10a).
Santo 1099(29a); 1099(29b); 3055(29a).
Santos & Barroso 2642(1a).
Sargent 10603(29).
Sassi Feb 1905 (19).
Sawyer 210(31).
Schade 151(6a).
Scheepers 901(14); 1120(3).
Scheepers & Terblanche A3(14).
Scheffler 252(27); 377(9).
Schelkownikow 324(29a).
Schell 1878 (29).
Schenk 5801(25).
Schery 5(29).
H.A. van der Schijff 2206(6b).
H.P. Schijff 3278(27).
Schimper 134(10a); 412(19); 542(9); 835(19); 837(17); 838(25); 838(9); 943(25); 1201(10b); 1419(9); 1424(9); 1755(9); 1755(9); sn (10b).
Schimpff 1058(9).
Schinizlein 1808(29a).
Schinz 29(1a); Jan 1886 (30).
Schlanzen 750(29).
Schlechter 183 C9(29b); 3647(1a); 3885(27); 4274(29); 6565(1a); 6684(14); 9911(1a); 11887(27); 11894(10b).

Schlieben 1794(9); 3302(27); 5772(27); 7266(1b); 7283(27); 7606(27); 8857(1); 8954(2); 10344(2); 10378(2).
Schlieben & Strey 8295(14); 8395(25).
Schmidt 575(2).
Schmiedel 364(1).
A. Schmitz 2354(6b); 3643(27); 7595(6b).
M. Schmitz 1489(1b); 4180(14); 6587(1).
Schneider 28 Aug 1882 (29a); 11 Sep 1918 (29); 9 Aug 1929 (31).
Schoenfelder S447(2); S618(6b).
Schofield 288(14).
Schrader 35(29a); 1810 (29a).
Schreber 365(1); 1767 (1); 1779 (1).
Schubert Apr 1965 (14).
Schultz 40(29b); 118(29b); 845(29).
F. Schultz 180(29).
Schunke 1964(29); 3011(6a).
Schur sn (31); sn (29).
Schweickerdt 1139(4); 1244(1b); 1284(1b).
Schweinfurth 11(10a); 93(10a); 105(29a); 108(29a); 111(29a); 115(10b); 116(10a); 118(10a); 143(10b); 336(10b); 392(19); 435(9); 458(29a); 549(29); 577(10a); 746(10b); 759(29); 900(19); 1026(19); 1088(10a); 2450(19); 2650(19); 3304(27); 10 Mar 1886 (10a); sn (19).
Schweinfurth & Riva 93(9); 184(9); 198(10b); 209(10b); 326(10b); 332(29b); 1036(9); 1828(17).
Schyff 2703(1a); 5877(14).
H. Scott & Britton 140(10a).
L. Scott Feb 1888 (29b).
G.F. Scott Elliot sn (18); 6534(18).
Sebastine 13108(29a).
Sedgewick & Bell 4852(31); 5367(29b).
Seemann 194(29a).
Semple 1703(1b).
Semsei 4006(19); 4034(19); 4037(29).
Sendevayi 243(1).
Seret 103(29); 811(27).
Seshagirittao Jul 1958 (29).
Seydel 46(30); 50(20); 83(2); 537(30); 1362(20); 1448(2); 1540(6b); 1541(2); 2275(20); 3439(6a); 4305(20).
Shabani 558(29a).
Shabetai 532(10a); 20 Jan 1933 (10b); 21 Jan 1933 (10a); 4 Apr 1944 (10a).
Shakya 9312(29).
Sharik S73-1698(29).
Shawani 26939(29a).
Sheldnick 1007(7).
Shell 13(31).
Shephew & Shephew 1 Jun 1938 (29).
Shimda 19 Jan 1974 (10a).
Shinners 8603(31).
Shiu Ying Hu 12131(29b).
Shuttleworth 99(6a).
Siame 127(6b).
Sichenrack 18 Dec 1897 (10a).
Sieber 366(6a).
Silk 154(2).
da Silva 1842(25); 2766(30); 3000(6b).
Simonds 26(10a).
Simony 17 Mar 1899 (10a); 18 Mar 1899 (10a).
Simpson 4381(10a); 4959(29a); 5045(29a); 5173(29a).
Simwandi 65(27).

G.C. Theron 152(1b); 719(4); 1630(2).
G.K. Theron 2968(6b).
J.J. Theron 526(14); 1247(14).
Thesiger 1 Jan 1946 (10a); 10 Feb 1946 (10a).
Thiebaud 790(6b).
A.S. Thomas 3443(29).
C.E. Thomas PI 124111(29a).
N.W. Thomas 1338(6a).
R.D. Thomas 86774(29b).
T. Thomas 1899(9).
J.B. Thompson 550(6a).
S.A. Thompson 320(6a).
G. Thomson sn (29b).
T. Thomson 50(29b); 344(29); 756(29a); 792(29a); Aug 1823 (31); Feb 1837 (10a); Sep 1846 (29a); 9 Aug 1849 (29b); 1849 (29b); 10 May 1850 (29b); 13 May 1850 (29b); sn (29a); sn (29b).
Thonar II 68(9).
Thonning & Guinen sn (29).
Thorrold 13 Apr 1932 (1a).
Thoset 41(29).
Thulin, Hunde & Tadesse 3405(10b); 3607(19); 3863(19).
Thulin & Mhoro 551(19); 2712(25).
Thwaites 3534(29); CV 3534(31).
Timberlake 2062(2).
Tinley 2542(6b).
Tinne & Tinne 47(25).
Tisserant 1976(25).
Tokyo University 22 Jul 1879 (29b).
Ton 3236(6a).
Torre 1905(2); 2115(14); 2643(29b); 2770(29b); 3756(27); 4613(27); 7893(1a); 8404(27); 8414(30).
Torre, Carvalho & Ladeira 18846(27).
Torre & Correia 15235(6b); 17561(6b); 17831(6b); 17837(27); 18177(25); 18359(14); 18366(14).
Torre & Paiva 9016(27); 9414(27).
Town clerk, Murrumburrah 2276/11(1).
Trapnell 452(14); 1767(6b); 1791(1).
Trauseld 1023(27).
Tropwell 1398(6b).
Trott, Dunn, Thurm, Case & Dziekanowski 175(6a).
Troupin 562(27); 600(27); 601(27); 689(27); 905(27); 1535(27); 4808(17); 5232(18); 5613(18); 6708(18); 7164(27); 7567(18); 8042(18); 8777(18); 11740(18); 11831(18).
Trujillo 8901(29).
Truman 23 Jul 1879 (31).
Trécul 1175(6a).
Tsiang 4560(31); 7158(29a).
Tucker 13235(23).
Tunin 6635(29); 6636(29a).
Turner sn (14); sn (29).
D. Turner 47(29).
Tweedie 1220(19); 2027(9); 2075(10b); 2269(19); 2274(19); 2327(10b); 4050(29a).
Tyson, Dwyer & Blum 3000(6a).
Tyson 1279(1); 1980(27); 2781(14).
Tölken & Hardy 628(20); 665(30); 678(30); 745(30); 790(20); 847(6b); 885(20).
Tölken & Schlieben 1170A(4); 1170B(1a).
Uotila 21080(31); 21103(31).
Vahameyer 1276(27); 1343(27); 2230(1).
van der Maesen 3902(29b).
van der Walt 5796(27).

Van Eseltine 1(29a); 3(29a); 5(29a); 6(29a); 9(29a); 10(25); 11(29a); 12(29a); 13(29); 14(31); 16(31); 17(31); 18(31); 19(25); 20(31); 21(31); 23(31); 24(31); 25(31); 26(31); 27(29a); 28(29b); 32(29); 38(29a); 53(29a); 9 Aug 1935 (29); 20 Jul 1936 (29); 28 Aug 1936 (29a); 31 Aug 1936 (29a); 5 Aug 1937 (31).
Van Meer 125(25).
van Rensburg 1668(27); 1716(27); 1857(27).
Van Royen 2075(2); 2175(2).
Van Someren 273(27).
Van Sow 28798(2).
van Wyk 6510(23).
Vander Gucht 227(29b); 410(31); 411(31).
Vanderyst 28023(6b); 28024(6b).
Vaughan 2254(7); 3068(17).
Venter 5086(27).
Veorteeg 1972(29b).
Verdcourt 1792(18); 1942(27); 2589(10b); 2621(17); 3177(10b); 3279(10b); 3897(19).
Verdcourt & Paulo 2160(18).
Verdcourt & Polhill 3164(17).
Verdoorn 1476(1b); Jun 1933 (1a).
Verheijen 2641(29b).
Verlot Aug 1870 (29a).
Versteeg 1972(29a).
Vesey-FitzGerald 1194(25).
Veterinary Inves Lab Nakurn M439(18).
Veterinary Officer CRS401(6b).
Vijay 39(29).
Viljoen 112(1a).
Viljoen 13(14).
Visagie Feb 1933 (1a).
Vogel Jul 1854 (29).
Volk 400(6b); 489(6); 819(30); 850(20); 2769(30); 2894(2); 2896(20); 2897(30); 11769(6b); 11918(6b); 12138(2); 12450(30); 12463(2); 12593(20); 12624(20); 12637(2).
Volkens 1957(17); 1972(18).
Vollesen 1962 (27); 3374(27); 3832(7); MRC2509(25).
von Rensburg 8599(27).
van Vuuren 1561(14); 512(14).
van Vuuren & Giess 1148(2).
Wager B188(2).
Wahab 1904 (10a).
Waitz 114(29).
Waller & Bauml 3083(29).
Wall 47(9); 1930 (29); 30 Oct 1936 (14); 20 Nov 1938 (14); 20 Dec 1938 (29).
Wallich 542(29a); 6734D(29); 1816 (29b); 1826 (29).
Walne 1807(29a).
C. Walter May 1892 (1b); May 1903 (1b).
H. Walter 4512(30).
H. Walter & Walter 534(6b); 1501(30); 1785(20); 2580(2); 2642(2); 2806(20); 3010(20).
van der Wal 12 Mar 1950 (28).
Wang 78475(31); 81969(32).
Ward Oct 1869 (10b).
C.J. Ward 1014(14); 2271(14); 3080(25).
E.N. Ward 8661/16(29).
Waring 73(29); 84(10a).
van Warmelo Mar 1960 (14).
Warming 20(6a); 88(6a); sn (31).
Waterhouse 330(29b).
G. Watt 7349(29); 2 Sep 1887 (29b).

2. Index to Scientifc Names

Accepted names are in roman type, and the main entry for each is in **boldface**. Synonyms are in *italics*.

Cucumis melo—Continued
  sortentypus *cassaba* 106
  sortentypus *chanaljac* 107
  sortentypus *dutma* 108
  var.-gr. *cantalupensis* 106
  var.-gr. *dudaim* 108
  var.-gr. *elongatus* 108
  var.-gr. *melitensis* 111
  var.-gr. *microcarpus* 111
  var.-gr. *reticulatus* 112
  ser. var. *adana* 105
  ser. var. *ameri* 105
  ser. var. *cantalupa* 106
  ser. var. *casaba* 106
  ser. var. *chandaliak* 107
  ser. var. *vard* 114
  ser. var. *zard* 114
  var. *acidulus* 104
  var. *adress* 105
  var. *aestivales* 105
  var. *aethiopicus* 105
  var. *agrestis* 81, 105, 115
  var. *albida* 105
  var. *alboviridis* 105
  var. *ananas-í-chair-verte* 105
  var. *anatolicus* 105
  var. *arundel* 106
  var. *atab* 106
  var. *aurantiacus* 106
  var. *autumnales* 106
  var. *azmirli* 106
  var. *baqubensis* 106
  var. *beida* 106
  var. *bitariana* 106
  var. *bos-valdy* 106
  var. *buharici* 106
  var. *bullock* 106
  var. *burrell's-gem* 106
  var. *cantalupensis* 106, 116
  var. *cantalupo* 106
  var. *cantonianus* 106
  var. *chandlkak* 107
  var. *chandulak* 107
  var. *chate* 107
  var. *chinensis* 107
  var. *chito* 107
  var. *conomon* 107
  var. *cossonianus* 107
  var. *cucurbitaceus* 107
  var. *culta* 107
  var. *cultus* 108
  var. *dudaim* 108
  var. *dumeri* 108
  var. *duripulposus* 108
  var. *elongata* 108
  var. *erythraeus* 108
  var. *firany* 108
  var. *flava* 108
  var. *flavus* 108
  var. *flexuosus* 109
  var. *fraiduni* 109
  var. *fucharici* 109
  var. *gem-rocky* 109
  var. *golobcresdael* 109
  var. *gourbek* 109

  var. *gracilior* 109
  var. *hafednafse* 109
  var. *hasanbey* 109
  var. *hibernus* 109
  var. *hiemalis* 109
  var. *hime* 109
  var. *honey-dew* 109
  var. *indica* 109
  var. *inodorus* 110
  var. *khadra* 110
  var. *kirukensis* 110
  var. *longus* 110
  var. *macro-castanus* 110
  var. *macro-leucus* 110
  var. *macro-pyrochrus* 110
  var. *maculatus* 110
  var. *major* 110
  var. *makua* 110
  var. *makuwa* 110, 111
  var. *maltensis* 110, 111
  var. *mansouri* 110
  var. *mehanawy* 111
  var. *micro-castanus* 111
  var. *micro-leucus* 111
  var. *micro-pyrochrus* 111
  var. *microcarpus* 111
  var. *microspermus* 111
  var. *minimus* 111
  var. *minutissimus* 111
  var. *momordica* 111
  var. *monoclinus* 111
  var. *oblongus* 111
  var. *ovatus* 112
  var. *persicodorus* 112
  var. *praecantalupa* 112
  var. *praecox* 112
  var. *pubescens* 112
  var. *reticulatus* 112
  var. *rigidus* 112
  var. *rokkiford* 112
  var. *rotundus* 113
  var. *rugosus* 113
  var. *saccharinus* 113
  var. *saharunporensis* 113
  var. *saidi* 113
  var. *samarrensis* 113
  var. *santawi* 113
  var. *senani* 113
  var. *shahd* 113
  var. *shammam* 113
  var. *shauki* 113
  var. *shimmam* 113
  var. *striata* 113
  var. *suavis* 113
  var. *tamago* 113
  var. *tarra* 113
  var. *texanus* 113
  var. *tuzensis* 114
  var. *typica* 111
  var. *utilissimus* 114
  var. *variegatus* 114
  var. *vaughans-original-osage* 114
  var. *virgatus* 114
  var. *viridis* 114
  var. *vulgaris* 114

Melo—Continued
  adzhur 115
    nid. *curvatus* 115
    nid. *subclavatus* 115
  agrestis 115
  X *ambiguua* 115
  ameri 115
    nid. *albidus* 115
    nid. *ananas* 115
    nid. *bargi* 115
    nid. *chiar* 115
    nid. *chtai* 115
    nid. *maculatus* 116
    nid. *wacharman* 116
  cantalupa 74, 78, 115, 116
  cantalupensis
    nid. *algeriensis* 116
    nid. *algeriensis* 116
    nid. *banana* 116
    nid. *carmelitanus* 116
    nid. *carmelitanus* 116
    nid. *compactus* 116
    nid. *compressus* 116
    nid. *primigenius* 116
    nid. *primigenius* 116
    nid. *turensis* 116
  cassaba 115, 116
    nid. *adress* 116
    nid. *aureus* 116
    nid. *burikala* 116
    nid. *candicans* 116
    nid. *gurgak* 116
    nid. *orientalis* 116
    nid. *zagara* 116
    nid. *zebrinus* 116
    nid. *zhukowskii* 114, 116
  chandalak 116
    nid. *bucharicus* 116
    nid. *caram* 116
    nid. *garma* 116
    nid. *glaucus* 116
    nid. *imdamas* 116
    nid. *loreus* 116
    nid. *tarnak* 116
    nid. *tochmi* 116
    nid. *variegatus* 116
    nid. *zamira* 116
    nid. *zami* 116
  chate 116
  chinensis 117
  conomon 117
  dudaim 117

  figari 103, 117
  flexuosus 117
  microcarpus 117
    nid. *albus* 117
    nid. *dudaim* 117
    nid. *erythraceus* 117
    nid. *flavus* 117
  monoclinus 111, 117
  orientalis 117
  persicus 114, 117
  sativus 117
  vulgaris 117
  zard 117
    nid. *asma* 117
    nid. *aurantiacus* 117
    nid. *gulabus* 117
    nid. *hibernus* 117
    nid. *kalassan* 117
    nid. *lapidosus* 117
    nid. *nokki* 117
    nid. *rugulosus* 117
    nid. *solidus* 117
    nid. *tarlama* 117
    nid. *tenebrosus* 117
Melothria
  pendula 90
Momordica
  charantia 89, 90
  cordifolia 89
  cymbalaria 91
  foetida 89
Mukia 19
  maderaspatanus 90
  micarantha 90
Myrmecosicyos 19
Oreosyce 5, 19
  africana 1, 89, 90, 91
  aspera 41
Orycteropus 78
  afer 78
Psiguria 91
  pedata 90
  trilobata 91
Raphidiocystis
  chrysocoma 89
Sicana
  odorifera 89, 90
Thladiantha
  nudiflora 89
Trichosanthes
  cucumerina 88

Addendum - *Cucumis* Monograph
Page 91
Cucumis sepium G.F.W. Meyer, Prim. fl. esseq.
278. 1818. = *Luffa sepium* (G.F.W. Meyer)
C. Jeffrey (Cogniaux & Harms, 1924; Heiser
et al., 1988; Jeffrey, 1992).
Page 96
Jeffrey, C. 1992. Names of the indigenous
neotropical species of *Luffa* Mill.
(Cucurbitaceae). *Kew Bull.* 47(4): 741-742.